sustainable
futures

sustainable futures

Teaching and learning: a case study approach

Edited by
Margaret Robertson

ACER Press

First published 2007
by ACER Press, an imprint of
Australian Council *for* Educational Research Ltd
19 Prospect Hill Road, Camberwell, Victoria, 3124

www.acerpress.com.au
sales@acer.edu.au

Edited by Emma Driver
Cover design by Italic Studio
Text design by Kerry Cooke, eggplant communications
Typeset by Kerry Cooke, eggplant communications
Printed in Australia by BPA Print Group

Acknowledgment
Extract from *In Siberia* by Colin Thubron, published by Chatto & Windus.
Reprinted by permission of The Random House Group Ltd.

National Library of Australia Cataloguing-in-Publication data:

Sustainable futures: teaching and learning: a case study approach.

For high school students.
ISBN 9780864314383 (pbk.).

1. Sustainable development - Case studies. 2. Environmental management -
Case studies. 3. Environmental education - Study and teaching.
I. Robertson, Margaret, 1948- .

363.7

Foreword

Anne Buttimer, University College Dublin

Past President of the International Geographical Union (IGU)

Rhetorics on global change abound. Imperatives for more sustainable ways of life weigh on human consciousness. Political voices proclaim the need for international cooperation on reducing carbon emissions; scientists call for interdisciplinary collaboration. For individual and local communities, however, the prospects of doing anything toward achieving these ideals in one's own immediate milieu remain opaque. *Sustainable Futures* offers potential solutions to this enigma via local, area-based initiatives by teachers and students. The book marks the harvest of a lengthy journey and an eventful maturation process. Hopefully it will also be catalyst for other ventures in many countries and schools throughout the world.

The initial seeds for this project were sown at a meeting of the International Council for Science (ICSU) in Paris in January 2001. There was much ado about environmental issues and the need for interdisciplinary collaboration, particularly on issues of humanity and planet earth. Questions of 'capacity-building' in science were also debated and a decision was made to grant each member scientific union a token sum of $5000 as seed money for educationally oriented activities. With a supplement of the same amount from the International Geographical Union's other 'parent' body, the International Social Science Council (ISSC), and encouraged by the welcome gift of a home base at Villa Celimontana by the City of Rome in 2000, the basic idea underlying this project was conceived.

It flowed from the conviction that geography, and particularly geography teaching in primary and secondary schools, could have a potentially vital role to play in raising awareness of environmental issues at all scales, from local to global. Capacity-building in science, given ICSU's declared commitments to 'society', should surely involve teachers and learners in precisely those courses where issues of global climate change and globalisation would most likely first occur, i.e. geography. Kofi Annan's plenary lecture to the Association of American Geographers meeting in New York 2001 already laid out the challenge for teachers of geography worldwide to avail themselves of data from the UN Global Geographic Data Base to raise awareness of issues in their own respective domains. So why not use these few dollars to engage teachers themselves in a pilot project aimed at facilitating local competence in the analysis and interpretation of locally based environmental questions?

Why not, indeed, just bring some geography teachers together at our newly acquired 'Home of Geography' in Rome and encourage each one to assemble a package of 'teaching materials' for courses on environment and sustainable development using data from their own respective home regions? The IGU Executive Committee agreed, and Vice-President Nikita Glazovsky, with characteristic generosity, volunteered to supervise this effort. Three teachers from three distinct cultural and linguistic worlds—Morris Chauke (South Africa), Álvaro Sánchez-Crispín (Mexico) and Nikolas Beruchashvili (Georgia)—bravely engaged in a research exchange network, first via email, then face-to-face at the Home of Geography for some weeks in autumn 2002. Together they designed a 'Teacher's Guide' for secondary-level courses on geography and sustainable development, illustrating this in the context of biodiversity and landscape, water and forestry resources, all of which were carefully documented with data from their respective home settings. By summer 2003 a first version of this text was ready.

The special IGU conference, *Society and environment interaction under conditions of global and regional changes*, held in Moscow and Barnaul in July 2003, afforded a good opportunity for Nikita and I to discuss this project with members of the IGU Commission on Geographical Education. We recognised that the central emphasis during the pilot round so far had been on substantive issues and that perhaps more attention was needed on pedagogical and didactic aspects. For these, as well as for future developments of the basic ideas, we looked to the Commission for guidance. It was at this point that Margaret Robertson, Commission Secretary at the time, graciously offered a hand and over the subsequent months she supervised a thorough revision of the language and structure of the text. Much enthusiasm was expressed by Commission members about further development of the project and its extension to a much wider range of participants. The final product from the pilot phase—possibly with translation to Spanish, French and Chinese—was anticipated on the occasion of the XXX IGC in Glasgow, August 2004.

Then in September 2003, after the meetings of the Institute of British Geographers in London, I had the great fortune of meeting Julia Marton-Lefèvre, former director of ICSU and then director of the Rockefeller Foundation's Leadership for Environment and Development (LEAD) program based at Imperial College. This was a highly encouraging encounter and we speculated on potentially joint projects of IGU and LEAD. Many of the most successful participants in the LEAD program in developing countries, she assured me, were actually geography graduates. And the Regional Director for LEAD activities in the Commonwealth of Independent States was none other than Nikita Glazovsky! Links were soon forged with the Chair of the British Council for Geographic Education, Ashley Kent, and ambitious plans were laid for joint contributions to the XXX IGC (Glasgow 2004) and for continued work with the IGU Commission on Geographical Education.

Meanwhile, at ICSU, the Priority Area Assessment on environment and sustainable development was acknowledging the value of place-based integrated research on human interactions with the physical environment—features so often proclaimed in geography but not always practised. The time seemed opportune to build on the experiences of the pilot phase of our project, extending efforts on the same themes to a larger number of contexts. With Margaret Robertson's energetic leadership, the competent collaboration of IGU Secretary-General Ronald Abler, and in consultation with a wide range of experts, soon a research proposal was submitted to ICSU. Entitled 'Geographical perspectives on sustainable development: Networking local area partnerships with teachers and young scientists', this mature phase of the project involved both innovative pedagogical strategies and community outreach, i.e. it sought to build local area networks with commitments to sustainable futures for natural protected areas, the use of forest resources and the management of water supplies. Reviewers of the project were impressed. 'The proposal is trans-disciplinary and ... if successful could be disseminated globally', the lead reviewer wrote. '[T]he project lends itself to being a model for ICSU to implement in other disciplinary and interdisciplinary areas.' The project was approved and, with supplementary financial support from IGU and especially from the Home of Geography, the work began.

Sustainable Futures: Teaching and learning: A case study approach offers details on this courageous journey. The process has no doubt changed the lives and perspectives of the participants themselves and presumably also those of the local area-based communities that engaged in the effort. Experiences of these past six years have yielded many a lesson, not always easy to learn! Challenges of cross-cultural and cross-disciplinary collaboration involve more than rhetorical statement or intellectual credo; they involve experiences of discovery, of not only some 'other' world of thought and action, but also the recognition that some taken-for-granted certainties and principles in one's own world pose obstacles to understanding other worlds. The technical possibilities of global communication via high-speed electronics are already within reach of individuals, but the extent to which they can facilitate mutual understanding among diverse worlds (intellectual, linguistic or cultural) or generate concerted action vis-à-vis the natural environment remains to be seen. There's still something which can be best achieved in face-to-face encounter and dialogue. Despite (or perhaps because of) their cultural and disciplinary backgrounds, participants in this endeavour really discovered a sense of community themselves. No doubt the cordiality experienced at Home of Geography, Margaret's untiring leadership and the thrill of discovering common denominators—as well as contrasts in taken-for-granted perspectives on nature and culture—all contributed to this sense of community. Director Giuliano Bellezza offered an apt name for the group: 'The Celiographers'.

Chairing a special session on this project at the IGU Regional Conference in Brisbane, in July 2006, I could only rejoice at the evidence that individuals had learned a great deal from their involvement. Several had already laid specific plans for further developments of the idea.

Clearly the challenge of global sustainable development calls for both 'top-down' global policies, democratically ratified by nation states on one hand, and at the same time 'bottom-up' well-networked local and regional initiatives which afford mutual inspiration and encouragement to others around the world. The Celiographers have dared to lay down stepping stones for this multifaceted challenge which could be tangibly workable in schools anywhere in the world.

Few were more aware of this prospect than the two individuals who were not, alas, present at the Brisbane event, Nikolas Beruchashvili and Nikita Glazovsky. Without their pioneering input, the pilot project might never have succeeded.

Contents

Acknowledgments   *xi*

Dedication   *xii*

Structure of the book   *xiii*

Contributors   *xv*

Part A: Challenges: Twenty-first century challenges associated
with sustainability, education and development   1

  Chapter 1: Introduction—Mobilising local and global networks   3
  Margaret Robertson and contributors

  Chapter 2: The gap between global rhetoric and local realities:
  How to build sustainable futures   22
  Margaret Robertson

  Chapter 3: Education for sustainability: Defining the territory   38
  Manuela Malheiro Ferreira

  Chapter 4: Sustainable development: What it means in the
  global context   49
  WU Shaohong

Part B: Case Studies: Eight chapters representing the diversity of
issues in global geographical locations   61

  The Asian region

  Chapter 5: Capacity-building for disaster management in
  southern Thailand   63
  Charlchai Tanavud

  Chapter 6: Water-saving in a city: A case study of Beijing City   79
  WU Shaohong et al.

Chapter 7: Rainwater harvesting in Mumbai   100
  Shyam Asolekar

Latin America

Chapter 8: Water resources, tourism and sustainable
development in the Tres Palos lagoon area, Mexico   126
  Álvaro Sánchez-Crispín

Chapter 9: Geography for urban sustainable development:
Students' proposals to deal with Santiago de Chile urban sprawl   146
  Hugo Romero and Alexis Vásquez

Chapter 10: Community and sustainable development:
Sustainability awareness through transportation, food and water
in a Buenos Aires neighbourhood   179
  Gabriel Fabián Bautista

Europe

Chapter 11: The forest fires issue in Portugal   191
  Manuela Malheiro Ferreira and Jorge Duarte

Chapter 12: Geographical perspective on training of students
in sustainable development in Georgia   209
  Niko Beruchashvili

Part C: Conclusion   225

Chapter 13: Our capacity for sustainable nation-building:
Do we flourish or flounder?   227
  Margaret Robertson

Appendix   *232*
Subject index   *233*
Author index   *237*

Acknowledgments

This book is indebted to the confidence placed in a small group of people who committed themselves to an ideal. Their vision has been to share the wonders of their homelands with learners of all ages around the world. The group fondly named 'The Celiographers' includes the authors of the various contributions in this book and many others. Encouraging the project throughout its many phases, International Geographical Union (IGU) President Professor Anne Buttimer has provided the confidence for participants to commit to the project. Working behind the scenes to secure the funding and arrangements for visas and travel, secretary-general of the IGU, Professor Ron Abler, has been a stalwart of support. The Director of the Home of Geography in Rome, Professor Giuliano Bellezza, and his colleagues, provided the space and quiet for moving forward the ambitious idea of this project into a practical set of activities. Organisations supporting the project include ICSU (International Council for Science), LEAD (Leadership for Environment and Development) and the Australian Academy of Science. In-country support has been received from numerous bodies, including the Russian Academy of Science and the Australian Embassy in Thailand.

Dedication

As authors we wish to dedicate our book to our colleague and friend Niko Beruchashvili. Admired and highly regarded by his colleagues in his homeland of Georgia and throughout Russia and central Europe, Niko brought a spirited independence to The Celiographers. Sadly, he passed away in March 2006. He was immensely proud of his work with our project and determined to see it through to a successful outcome. Fortunately for us he had completed his case study and submitted the written manuscript which we publish with the collection. *Spasibo*, Niko.

Structure of the book

Our book provides an illustration of a working model for meeting global targets at local levels. The accepted need to *think globally* and *act locally* is an implicit understanding. Involving scientists from developing nations in nine different countries located on six different continents, this collaborative work provides an inspirational record of the levels of cooperation and shared vision that can come from talking and working together. Language has always posed an inordinate barrier to global communication, as have the associated norms and values embedded in the national and localised cultures. The first hurdle in the process of coming together with a shared set of purposes is conversation and a willingness to listen to the voices that represent local views. The richness of the dialogue and the ensuing sense of shared ownership of the outcomes has been a truly remarkable journey for all who have participated. The stories of these experiences through case studies of local projects aimed at environmental management and education for better futures provide a small working model of the global community at work. Broadening horizons of young people involved in local projects provides the environmental education and 'know-how' needed to support locally driven needs.

The book is organised in two major sections. Part A consists of four chapters. Each presents a different element of the quest for improved understandings related to the process of teaching and learning for sustainable development. We hope that each of the chapters will inform and lead to critical enquiry in formal learning environments and well beyond into the community of educators. Chapter 1 provides the background information to the origins of case studies presented in Part B. It contains useful information on how to develop communities of practice that can sustain cross-cultural, linguistic and geopolitical differences. For those who are sceptical about the possibility of shared dialogue against the many obstacles to be encountered, the process outlined should provide a useful model for future projects.

Chapter 2 reaches into the soul of people–place relationships and the need for education. References are made to the inspirational leaders of the twentieth century who managed to guide their people towards shared visions of sustainable lifestyles. Strategies outlined include the construct of Communities of Practice and Activity Theory.

In Chapter 3, Manuela Ferreira reviews the common and agreed international meanings of sustainable development. Reference is made to the international agreements that include Agenda 21. A useful application of the ideas to the European context completes the chapter.

Following on from the definitions developed in Chapter 3, WU Shaohong completes this first section of the book with an overview of the international state of the world's major natural resources.

Part B consists of eight chapters related to the country case studies. In order, including themes, they are Chapter 5, Thailand (disaster management); Chapter 6, China (water management in Beijing); Chapter 7, India (rain water harvesting); Chapter 8, Mexico (ecotourism); Chapter 9, Chile (urban water management in Santiago); Chapter 10, Argentina (community sustainability in Buenos Aires); Chapter 11, Portugal (forest fire management) and Chapter 12, Georgia (teaching for sustainability in Georgia).

The final section, Part C, contains reflective comments on the case studies raised, with particular reference to the case study for South Africa.

Contributors

Dr Shyam R. Asolekar (India) is Professor at the Centre for Environmental Science and Engineering at the Indian Institute of Technology. Dr Asolekar's primary interest lies in teaching undergraduate and postgraduate students, as well as industrial and regulatory professionals.

Nicholas Beruchashvili (Georgia—deceased) was born in 1947 in Tbilisi, and was Professor Chair of Cartography and Geoinformation at Tbilisi State University. He was an active researcher in many European countries, and spent periods of time in Paris, Warsaw, Sofia, Brno, Toulouse and Grenoble.

Anne Buttimer (Ireland) has been Professor Emeritus of Geography at University College Dublin since 1991, and was President of the International Geographical Union, 2000–04. She has held research and teaching positions in Belgium, Canada, France, Scotland, Sweden and the USA.

Morris Chauke (South Africa) is a school principal at Tshikhwani Combined School in Limpopo Province of South Africa. He pioneered South Africa's participation in the International Geography Olympiad for Young Geographers.

Manuela Malheiro Ferreira (Portugal) is Professor at the Department of Sciences of Education, Universidade Aberta, Lisbon, and a researcher at the Centre for Studies of Migrations and Intercultural Relations. She has written widely on geography education, intercultural education, citizenship and sustainable development education.

Margaret Robertson (Australia) is the Executive Secretary of the IGU's Commission for Geographical Education and currently Associate Dean of Research in the Faculty of Education at La Trobe University, Melbourne. She has a career-long interest in visual–spatial pedagogy.

Hugo Romero (Chile) is Professor in the Department of Geography and Head of the Interdisciplinary Program on Environmental Planning and Management in the Universidad de Chile. His research interests are mainly in urban and regional sustainable development.

Álvaro Sánchez-Crispín (Mexico) is a researcher based at the Institute of Geography of the National University of Mexico (UNAM), Mexico City. His research interests are concerned with economic geography issues, particularly the geography of tourism and the economic assimilation of space.

Charlchai Tanavud (Thailand) is Reader in Soil Science at the Faculty of Natural Resources, Prince of Songkhla University. Current interests include the use of geographic information systems and remotely sensed data in natural resources and environmental management, and disaster preparedness.

WU Shaohong (China) is a professor specialising in physical geography and environment and resources at the Institute of Geography of the Chinese Academy of Sciences. WU's interests in sustainable development education focus on global change and its impact on natural ecosystems, and how natural ecosystems respond to global change.

Part A: Challenges

Twenty-first century challenges associated with sustainability, education and development

1 | Introduction—Mobilising local and global networks

Margaret Robertson and contributors

Global concerns for the future of our planet spark many emotions and responses. Our book is no exception. However, what is recorded in the chapters that follow goes well beyond the rhetoric of scientists, government officials and concerned observers. We accept the reality of the grave situation facing the global village and provide a series of explanations of how local communities can build their resources and capacity to combat the problems. Our focus is twofold: first, young people located in local villages and neighbourhoods of developing nations. They share a space where there is limited access to education and often even less access to the basics of survival including fresh water, adequate sanitation, and provisions for hygiene and health. Second, we wanted to develop a set of authentic education resources on sustainable development initiatives for use in schools and higher education around the world. Our shared vision conveyed through the chapters of this book contains a story of immense courage and imagination that should challenge any community regardless of its context and circumstances.

Inspired by the work of three scientists who collaborated to develop case studies of their sustainable development projects in Mexico, Georgia and South Africa respectively,[1] the stories recorded in our book expand this initiative to include nine further case studies. Representing every continent, the scientists who worked on this project provide an inspirational model of how to take the rhetoric of sustainable environmental management into the everyday lives of people in remote and fragile regions of the world. This outcome has come about from a remarkable set of events.

The steps that preceded the case studies reported in our book are worthy of reflection for future projects. They demonstrate how to take a kernel of an idea and bring it to fruition as a practical educational outcome, and they unfolded in the following sequence:

- An international funding opportunity presented itself for project initiatives aimed at educational capacity-building for young scientists concerned about environmental issues in developing countries.[2] Support for the project was also forthcoming from Leadership for Environment and Development (LEAD) which is an independent, not-for-profit organisation.[3]

- An application for project funding was developed in 2004, premised on three major principles:
  — The project proposal placed a high value on developing local area networks of skilled people who could lead manageable, small-scale projects.
  — Leadership at national levels was important for maintaining the local case study momentum and for continuing activity including further projects.
  — National leaders were to be part of an international cooperation with a shared vision of how the projects needed to unfold. They would also share common principles for cross-cultural comparisons and gaining maximum benefit from the dissemination of the reports.

- Specific aims for the project were to engage teachers and young scientists in developing countries in science projects aimed at major contemporary sustainable development issues, including biodiversity, forestry and water resources. The project worked from the principle that sustained outcomes are linked with community ownership. Hence, it relied on negotiation, community partnerships and recognition of project outcomes through existing and newly created structures. Capacity-building, in our view, required us to think more broadly about the meaning of sustainability and environment to include local wisdom about environmental events and processes.

In Chapter 5, Charlchai Tanavud illustrates this issue in relation to disaster management of tsunami-related flooding. While the scientific data may provide early warning signs of pending catastrophes, changed behaviours in the fauna and flora can also signal responses to unusual atmospheric patterns. Interestingly, at the time of writing, the El Niño effect is causing significant disturbance to the weather patterns in the eastern Pacific Ocean. Resultant droughts in some areas and floods in others are affecting all aspects of animal and plant life. Phenomena such as El Niño and tsunamis, as we have witnessed most recently in South-East Asia, can result in catastrophic loss of human life. While we tend to reflect on such natural phenomena as being unusual and experienced irregularly across a time span of human life, their occurrence also highlights the fragility of our management practices and capacity for survival on planet earth. The continual monitoring of our land use responses can react to events as they unfold, or we can better prepare for these events using scientific resources that can indicate the need to change our human responses to the landscape.

- Related to sustainable development, the *expected project outcomes* were:
  — *local area networks* with commitment to sustainable futures for biodiversity and natural protected areas, the use of forest resources and water supplies

— *expanded networks* for communication of project results with local, national and global communities
— expanded *intercultural learning* and understanding of related local issues
— scope to identify *future leaders* in our scientific community and encourage their quest for knowledge and skills
— a *global mechanism* for ongoing collaboration with partners in developing countries with the potential for expansion to other countries and new partners.

The funding application received positive feedback and was successfully funded. We then had the much-needed financial boost to support the ideals embedded in the project aims. A brief outline of the project process follows. Our view is that the process is worthy of recording in this book as a guide for reflection on how to bridge the gap between local, national and global interests.

1. The project process—turning the idea into practical outcomes

As we would for any sound educational project, we worked towards a predetermined set of educational and research objectives. The context was our first concern. However, producing authentic teaching and learning projects that could be used anywhere in the world posed its own inherent problems. At the most basic level, we needed a common language for developing a shared dialogue and debating concerns. This is never an easy process. Predictably, the language shared by all participants was English, but English is not the first or even second language of some of the participants; in some cases it is the third or fourth language. Although we acknowledged this as discriminatory, we worked from the assumption that our work required the scaffolding of skilled in-country scientists who could make the translations and reach out to local people. Hence, English is the enabling tool we used for our planning and sharing meetings in a common space, for communication during face-to-face meetings, and through continuing online discussions. The global initiatives are in turn translated to local languages for broad application and transmission of ideas. The other important issue to consider was that we are all volunteers working with limited budgets and time. Our dedication had to come from a sincere will to make this project work.

The following set of generic objectives reflected the perceived global imperatives of the project. They included the need to:

• shift from the scientific to the *applied side of geography*. That is, we needed to recognise the science–place interface.
• recognise that *policy is place-specific*. We needed to acknowledge that what works well in one place will not work in another. Cultural specificity is always an important dimension when implementing globally driven projects.

- recognise that *acting locally* and thinking globally underpinned the pedagogical approach
- *evaluate* previous policies to determine the project parameters. This was the background knowledge we required to assess needs.
- bring the rhetoric to local levels and individual commitment through building strong *communities of practice*
- encourage *local 'discovery'* of environmental awareness
- promote knowledge, through resultant local networks, of the *environmental 'commandments'* for the global community.

An **eight-step implementation** process has taken these ideas from a set of broad objectives forward into practical outcomes. While the steps themselves are specific to this project, they provide a **process model for project management** that can be applied to multiple contexts. Hence we record the essential information of these steps as a case study model for future projects, highlighting the obvious aspects to consider as well as unexpected bonuses.

Step 1: In-country scholarship process

In-country working parties were set up to act as selection panels and provide support for project scholarship recipients. Scholarship recipients were local scientists and educational leaders with national and international connections, who were recognised as key people to make the connections from the global levels to the local comunities and villages of their people. Nine people participated, representing the following countries: the Asian region—China, Thailand and India; Europe—Portugal and Georgia; Latin America—Mexico, Argentina and Chile; Africa—South Africa. Representing the other major continent of Australia, I had the privilege of managing the project. Each of the participants is a co-author of this book; their projects are recorded in separate chapters of the book.

Step 2: Professional learning and project planning—Rome period

Participants prepared proposals for their local projects for an intensive, two-week, face-to-face planning phase and professional learning program held in Rome at the Home of Geography[4] during June 2005. During this time participants worked closely together, using collegial approaches, to design a project ready for implementation within a three months after returning to their home country. Developing trust was an essential part of this period of being together. We needed to build a strong sense of belonging to maintain the project momentum once back in our respective countries.

Developing a group identity was part of this process. Within this two-week space of time the group formed and launched its identity as 'The Celiographers' (the geographers of our temporary home at the Villa Celimontana—also the sprititual home of the project). Working on agreed principles for project design, plans were finalised and circulated for final endorsement by the newly formed group. Consensus by collegial debate was an important dimension during this phase.

Figure 1.1   The Celiographers at work

Step 3: Local area training

The Celiographers worked on developing local area projects, which are presented in the chapters that follow. Summary outlines of all three projects are contained in the section that follows this eight-step process. The details of these are being reported on the project website (see Step 6).

In-country projects adopted a similar model to the one used in Rome by the international team. Each received a small funding budget from the project. Guided by the in-country Project Leader and acting through Local Area Network Leaders

(LANLs), this phase required working collaboratively with local educational and governmental officials to train local groups of teachers and/or interested (science) students. LANLs were supported by a corresponding mentoring program including access to resources and relevant expertise. For the project to work, and for the planned increase in capacity to occur, this transmission of knowledge and skills to able students and teachers was crucial. The in-country Project Leader was expected to recognise and, if at all possible, overcome local obstacles. In all countries but one this was achieved. To provide advice and support, and to share outcomes from other projects, it is expected that at mutually agreed times both the educational and scientific experts will work with the LANLs within their countries.

Step 4: Developing the local area projects

In order to develop locally modified models of best science teaching practice, local teachers/students were invited to apply their new knowledge in a monitored local area project. Projects were based on agreed pedagogical principles of negotiated decision-making and collaborative community partnerships. There needed to be clear expectations of what the project was about; what it hoped to achieve; how it would meet its targets; what the likely obstacles were and how these might be overcome; how to motivate local colleagues, friends and community members; the local strengths, including wisdom and sources of advice; and what to observe, record and evaluate.

The construct of 'Action Learning', or reflective action, was a strong component in the project team thinking. We worked along the lines shown in the equation below, adopted from Revans (Robertson, Webb & Fluck 2007). The assumption is that we all have an existing body of knowledge, or programmed knowledge, which enables us to develop insights and new understanding through questions that probe existing understandings and develop new meanings.

Programmed knowledge + Questioning insight

=

Learning for meaning

Step 5: Project evaluation and assessment

Each of the chapters that follow reveals aspects of project evaluation. These dimensions vary and serve to highlight the constraints that inevitably occur when you move into the realm of everyday living. At the international level there is widespread agreement that the project model is working well and much has been learned. A few of its achievements are:

- broadening networks of scientists and educators working on shared problems
- transformed practices in local behaviours
- enhanced self-confidence leading to capacity-building that can spiral out to increasing numbers of committed and well-equipped people
- better understandings of local problems and locally derived solutions
- a bank of much-needed case study resource materials for the study of environmental issues and locally driven solutions
- the development of a mechanism for ongoing coordination and collaborative partnerships
- new possibilities for exchange experiences for scientists, students and teachers.

Step 6: Sharing the knowledge

Local

In line with the project's aims, sharing the knowledge was part of the brief for The Celiographers. In-country project teams have been encouraged to provide opportunities for local area projects to be showcased and shared with wider communities. Such occasions enable participants to learn from one another, and build on their initial contacts for future projects. Identifying potential curriculum initiatives in schools that articulate with the project goals for sustainable development has been a major aim of this meeting phase.

International

The Celiographers have reported their findings at international symposia and maintain an active email network to sustain their efforts for future presentations to the international scientific community.

Publication output

This book is the first major publication of the group but by no means the only print outcome. It is expected that reports from all projects will be made available to the scientific community through a website. Ongoing publications will be part of the next generation of targeted funding to introduce new participants and share our experiences with a widening network.

The website (see http://www.celiographers.org)

This site has been developed and maintained by volunteer members of The Celiographers in conjunction with the support of the International Geographical Union. The website provides opportunities for project leaders to discuss ideas and network widely. It also provides the platform for disseminating information on regional sites via local languages. This public forum adds to the existing knowledge outputs through journals and scientific reports.

Step 7: Project evaluation and report writing—only part of the picture

The project's outcomes, and how effectively it meets agreed targets and goals, depend ultimately on the timely goodwill and opportunities that occur where the action is to take place. This project has been no different from any other in this respect. Accepting the reality of everyday life as our ultimate judge of performance should not lead us to see too much imperfection. As Pile and Thrift (1995) observe: 'The human subject is difficult to map ... There is difficulty in mapping something that does not have precise boundaries ... that is always on the move, culturally, and in fact' (p. 1). For our project team, the moving context has added to the flexibility in responses that we see as essential for success. We have learnt to appreciate each other's strengths and work as a team. Our collective view is that we can build on *making a difference* in people's lives, regardless of how small this progress may at first seem. Human agency cannot be judged by absolute conformity but by the quality of the dynamic it can generate. Striving for a community of practice, such as The Celiographers have done, has intrinsic rewards, which all participating individuals carry within them regardless of externally observed evidence. Social capital is less observable or tangible at first inspection.

Step 8: Maintaining the momentum of change—developing the social capital

Any change process is fragile. Good leadership can make a significant difference for transforming a task into a lasting, lived experience. Developing the points made in Step 7, Roseland (2005) makes the point that success with sustainable communities requires knowledge of 'how to *multiply social capital* for sustainable community development' (p. 10). In Chapter 2 some of the issues related to lasting changes and transformative practices are considered in some detail. The emphasis is placed on agency and developing a set of ethical beliefs that can sustain principled commitment—if not specifically to the project reported in the case studies of this book, then to the fundamental principles related to our shared responsibility as a global community to custodial management of planet earth.

In summary, these 8 steps or phases have provided the basis for the projects described in the chapters of this book. They provide a guide for future projects and a model for collaborative planning across boundaries and seemingly extremely complex odds. Setting aside issues related to visas and financial support for the international participants each project reveals aspects of local governance and agency that have both simultaneously helped and hindered the in-country projects. Our collective belief is that end result is testimony to the courage and determination of the individuals involved. Their stories follow.

2. Place–space connections—the scientific aspects at a local level

The objectives for each of the in-country projects worked from the broad principles outlined in the beginning of this chapter to the specifics of making the plans happen. Project leaders shared a commitment to geospatial knowledge, hence maps and visuals imagery of the case study sites are used throughout the book to capture the local issues. Additional multimedia material is envisaged for the next phase of the project. Intercultural learning about other people and their places and spaces should also come through the stories. Resultant enquiry at local levels in schools and higher education institutions could well focus on critical thinking about available resources and infrastructure, local skills and knowledge, and the more descriptive aspects of the place and its people, including culture and beliefs.

The enquiry framework used for all the projects was the basis of the Rome training period. As the template in Figure 1.2 (pages 12–13) indicates, that emphasis was placed on the learning process. In general terms, this framework helped shape each in-country project so that a level of coherence in the reporting has been possible. Teachers should find the template useful as a guide to the projects described and for translating to their local areas.

3. Introducing the in-country projects

The Celiographers share a deep concern for planet earth and for the wellbeing of humanity. In the following introductory statements, each member of our group outlines their concerns and priorities. Their perspectives provide an introduction to the beliefs that underpin our collective thinking on sustainable development issues. Subsequent chapters in Part A provide expanded information of what we mean by 'sustainable development' and related geopolitical issues.

The Asian region

Thailand—Charlchai Tanavud
Sustainability has been widely discussed in resources management. While the definition of sustainability depends on the context within which it is being used, it is generally accepted that the paradigm of sustainable development requires a positive relationship to be created between economic growth and environmental harmony. In southern Thailand, economic growth, environmental quality, employment and local participation are essential components for sustainable development.

Over the years, in response to necessity and economic opportunity, much of the forest resource on steep upland slopes of southern Thailand has been gradually

Subproject 1: Title

Project country / region / locality

A description of the place context for the project.

Resources

Summary of existing and possible sources of local support and publicity, for example:
- government—local, regional and national
- educational
- non-government
- other resources.

Aims and objectives

Please specify clearly how you see project aims translated to your own national, regional and local contexts.
Aims:
Objectives: written as active outcomes, i.e. using *verbs*, for example:
Knowledge:
1   Students will *gain* knowledge of the geographical content of the local area.
2   Students will *gain* knowledge of the water problems in Beijing.
3   Students will *investigate* available resources on water issues in Beijing.
Skills:
1   Students will *develop* mapping skills.
2   Students will actively *cooperate* in group work.
3   Students will **develop** questionnaires related to the water problem.
Values:
1   Students will *increase their understanding* of the problems of sustainable water supply.
2   Students will *develop appreciation* of the problems.
3   Students will *appreciate* other perspectives or points of view.

Project description

Outline the local social context for each in-country project, including local customs, values and beliefs; likely key people to engage or involve in the projects; likely obstacles; and possible ways to overcome these obstacles. Include the following:
- target group (the learners)
- project aims and outcomes (draft only)
- key people (stakeholders, e.g. schools, NGOs, local authorities, university students)
- local political context relevant to the project's success
- current skills and available resources
- skills needed and why
- anticipated obstacles
- other resources funding that may be available
- other relevant information.

Figure 1.2   Common template used for designing projects

Sustainable Futures Copyright © Margaret Robertson 2007

Possible learning experiences / activities

Select and describe in detail how the learning experiences will facilitate the learning.
Tick ✓ *and* add to list:

Listening ☐
Talking ☐
Observing ☐
Group work ☐
Problem-solving ☐
Other ☐

Draft task feedback proforma

Tasks, including criteria, for example research skills, written communication skills, analytical skills, organisational skills, knowledge of the subject, appreciation of the problem, etc.

Criteria	Excellent	Well done	Satisfactory	Unsatisfactory
1				
2				
3				

Individual project evaluations

Evidence gathered and criteria for judgments, for example research skills, written communication skills, oral presentation, analytical skills, organisational skills, knowledge of the subject, appreciation of the problem, etc.

Criteria	Excellent	Well done	Satisfactory	Unsatisfactory
1				
2				
3				

Evaluation

What evidence do you have to show that the teaching outcomes were:
- successful?
- partly successful?
- less than successful?

Figure 1.2   Common template used for designing projects (continued)

Sustainable Futures Copyright © Margaret Robertson 2007

replaced by rubber plantation. The encroachment and destruction of this resource has led to the occurrence of accelerated soil erosion, with consequent impacts on soil productivity and environmental quality. Land use planning should, therefore, be formulated to facilitate the optimal utilisation of the land resources while maintaining and/or improving environmental quality. It is anticipated that the implementation of sustainable land use planning, together with an array of actions including employment generation and people participation, would give substance to the general concept of sustainable development that should guide future development of southern Thailand.

The catastrophic tsunami that occurred in southern Thailand on 26 December 2004 has been described as the worst natural disaster in Thailand's recorded history. These giant waves claimed some 5000 lives and destroyed or severely damaged hotel resorts, tourist facilities and infrastructure. Such losses have deprived the southern part of the country of important resources which could otherwise have been used for economic and social development, thereby impeding the process of sustainable development. It is anticipated that adopting effective preparedness options to supplement existing measures to reduce disaster risks would enable the southern region to become more resilient to the effects of tsunami hazard, thereby supporting sustainable development in southern Thailand.

China—WU Shaohong

The development of any region should not compromise the interests of any other region. Development should be based on the principle of equality—between generations of people, regions and sectors. A balance between the interests of people and of nature also needs to be struck. China's population growth in its large cities places extreme pressure on available resources. Development of sensitivity to reduce urban sprawl and to use public transportation, along with an awareness campaign for families, especially the young generation to use resources responsibly, is part of the scientific strategy for the future.

The key words in the China case study are:

- sustainable development
- water resources shortage
- water saving
- daily life
- green belts planting
- commercial activities.
  The key questions for this case study are:
- Should we take care of natural resources (water resources)?
- Could we save natural resources (water resources) by changing some of our behaviours?

Editor's note: These ideas are further expanded in Chapters 4 and 6.

India—Shyam Asolekar

The following declaration was made on 20 January 2005, by more than 800 learners, thinkers and practitioners from over 40 countries, engaged in education for sustainable development, at the *Education for a sustainable future* conference held at the Centre for Environment Education, Ahmedabad, India.

The Ahmedabad Declaration

As the first international gathering of the United Nations Decade of Education for Sustainable Development (DESD), we warmly welcome this Decade that highlights the potential of action education to move people towards sustainable lifestyles and policies. If the world's peoples are to enjoy a high quality of life, we must move quickly toward a sustainable future. Although most indicators point away from sustainability, growing grassroots efforts worldwide are taking on the enormous task of changing this trend. We accept our responsibility and we urge all people to join us in doing all we can to pursue the principles of the Decade with humility, inclusivity, and a strong sense of humanity. We invite wide participation through networks, partnerships, and institutions.

As we gather in the city where Mahatma Gandhi lived and worked, we remember his words: 'Education for life; education through life; education throughout life'. These words underscore our commitment to the ideal of education that is participatory and lifelong.

We firmly believe that a key to sustainable development is the empowerment of all people, according to the principles of equity and social justice, and that a key to such empowerment is action-oriented education. ESD [Education for Sustainable Development] implies a shift from viewing education as a delivery mechanism, to the recognition that we are all learners as well as teachers. ESD must happen in villages and cities, schools and universities, corporate offices and assembly lines, and in the offices of ministers and civil servants. All must struggle with how to live and work in a way that protects the environment, advances social justice, and promotes economic fairness for present and future generations. We must learn how to resolve conflicts, create a caring society, and live in peace.

ESD must start with examining our own lifestyles and our willingness to model and advance sustainability in our communities. We pledge to share our diverse experiences and collective knowledge to refine the vision of sustainability while continually expanding its practice. Through our actions we will add substance and vigour to the UNDESD processes. We are optimistic that the objectives of the Decade will be realized and move forward from Ahmedabad in a spirit of urgency, commitment, hope, and enthusiasm. (CEE 2005)

Latin America

Mexico—Álvaro Sánchez-Crispín

While many people would relate the issue of sustainable development to the realm of natural sciences, the true origin of this idea is deeply rooted in the social sciences, among which geography stands prominently.

Sustainable development is the process through which decisions about the way we do things, individually and collectively, locally and globally, that will enrich the quality of life now, will not damage the planet for the future. In other words, sustainable development is a course of action in which the capacity of the environment (natural and socioeconomic) to meet present and future needs is not compromised. Thus, natural and socioeconomic issues such as poverty, equity, quality of life and global environmental protection, among others, are part of the sustainable development debate. Central to this debate is the question of *balance* between the decision-making of today, with regard to the fragile nature–society relationship, and the needs of future generations.

Sustainable development can be examined from different perspectives and by different sciences. However, as pointed out by Wilbanks (1994), the connection between geography and sustainable development can be demonstrated from four distinctive points of view:

1   *Diversity*. This includes both natural and socioeconomic diversity, from the polar regions to the tropics, from the urban environs to rural communities, from sophisticated technologically advanced societies to the most isolated human groups on earth, from terra firma to marine milieux, from traditional agricultural lands to heavily industrialised regions.
2   *Nature–society, temporal and spatial flows*. This point makes distinctive the geographical analysis of sustainable development from other type of scrutiny. What is important here is to grab hold of the territorial component of sustainable development.
3   *Questions of scale*. This is a fundamental part of the geographical analysis as it gives way to generate a crystal-clear picture of how sustainable development could be achieved at local, regional, national and planetary levels.
4   *Power of visual images in the information revolution*. Associated cartographical production is central to any geographical analysis. In view of this, spatial and territorial features of sustainable development can be charted for different purposes, including educational, political and planning.

Chile—Hugo Romero and Alexis Vásquez

Sustainable development from a geographical point of view requires setting up spatially well-defined and decentralised institutions and regulations that ensure the auto-realisation of people, according to their expectations, conditioned by local identities, and following their own social and cultural values.

Local societies must administer natural resources, and environmental goods and services, in such a way that they can remain permanent (like natural capital) and be used according to their resilience. Material sustainability should be guaranteed by local, regional, national and international authorities, but at the same time the natural resources basis should be protected, and there should be enough capacity to confront hazards, uncertainties and disturbances. Natural resources and the conservation of nature require the existence and location of protected spaces to perform their environmental functions and services, and special measures and production practices to be sustainable in the long term.

Economic benefits obtained from natural resources, industries and services should be locally retained by places and regions, and democratically shared by the society, ensuring adequate levels of socioeconomic equity for all members. Economic progress, social equity and the promotion and conservation of cultural values must be permanently internalised by the people and their institutions. Poverty, social disparities and the loss of local cultures and identities must be avoided. Education, health care and access to information are some of the most important services that governments and local authorities must ensure for the total population.

Social networks should be strengthened to allow collective commitments, social protection and a safe and peaceful existence. Equal accessibility of all the people to markets, service centres and recreational facilities should be guaranteed by the government and respected by society. The respect of human rights is a priority issue and a social aim, shared by all the members of society. Access to justice, a free press and public information is a constitutive part of democracy. Social integration and spatial convergence are facilitated by the creation and development of places where central services are offered, such as local markets, public education, parks and public spaces, connecting roads, and social and cultural facilities. Social and cultural integration are common aspirations, which avoid social exclusion and socio-spatial segregation.

Under globalisation, one of the most critical issues is economical, cultural and political asymmetry between local and regional communities and world-scale private transnational companies, financial institutions and large cultural systems. The setting up of institutions and regulation that improve the quality of the spatial relationships and the symmetry of global–local exchanges seems to be essential in the present world, where differences between developed and developing nations or regions have increased.

Argentina—Gabriel Fabián Bautista

Sustainable development is usually meant to be related to the future, especially to the future generations. It is a moral issue because it comprehends an intergenerational equity; yet it is mainly moral because it is an intra-generational equality as well; that is, it is about the present. It is at this moment of the beginning of the twenty-first century that there is a disproportion between lifestyles in the north and in the south, and a general crisis for both. Disproportionate access to resources constrains the opportunity of millions to develop their own human vocation to happiness and meaning. By addressing this, the earth becomes a home for everyone instead of a wasteland, a home for building our own capacities to make things happen according to projects coherent with our own cultures and perspectives.

The geography of sustainability means to construct the place as a home and pass it over to the future generations. This sense of place is evocative of the long duration of human works and gives a sense of intimacy, not alienation. The global is the horizon, but this does not blur the local. The local is the urban environment as much as the rural and the natural. Maybe the urban and the rural are more important than the natural because almost all the world's population lives in urban and rural environments. It is a complex whole, that involves risks and a sense of limits within the second nature as the first nature is reconfigured through human activity. It means the re-inhabitation of the land, an awareness of the moral condition of liberty.

Sustainability is not only about the appropriate use of technology and appropriate management of natural resources. It is about culture and world views. Every nation has the right to employ its own resources, but this must be done with consideration for the common good; so developed nations, too, should reconfigure their own economies and re-educate to be able to willingly renounce the addiction to growth in terms of gross domestic product.

Sustainability is not a utilitarian technological procedure; it looks forward to the development of the whole human person and every human being on earth. It is a relationship with the land, and with the others as community and with a transcendental vision of life, which is not constrained to the present condition but open to the future and the utopia of a new earth that will come as a gift. The awareness of the earth and life as a gift is essential to avoid the purely utilitarian logic to control and command; this opens up a more humanistic management and a world of meaning and values. Otherwise development is merely an accounting of resources, and a maximisation of profits in the short terms.

Europe

Georgia—Niko Beruchashvili (deceased)

Questions of sustainable development should be related to the 27 principles established at the World Summit in Rio in 1992 (Agenda 21), or at the Johannesburg World Summit in 2002, or to other authorities on sustainable development.

In our project, *Geographical perspectives on sustainable development: Networking local area partnerships with teachers and young scientists*, it is important to explain to students:

- how we can manage processes in the nature and a society
- that this management should be focused on sustainable development
- that sustainable development is development that meets the needs of the present without compromising the ability of future generations to meet their own needs

- how all possible variants of sustainable development are connected to spatial or spatial–temporal development of the country (territories, regions, local societies)
- that sustainable development conflicts with nature, or with social needs of a society, or with other countries (other territories), or with itself (other variants of development)
- that there are different variants of development; one of these variants brings economic gain, and others bring further economic gain. The main thing that sustainable development should provide is a normal life not only for our generation, but also our children's generation.

Problems need to be posed so that students understand the variants of development. Teachers need to encourage pupils to:

- understand the essence of a problem of sustainable development
- understand the complexities of development
- analyse the positive and negative sides of the arguments
- argue their opinion (on the basis of the evidence) or their point of view on sustainable development.

Portugal—Manuela Malheiro Ferreira

'Education for Sustainable Development is fundamentally about values, with respect at the centre: respect for others, including those of present and future generations, for difference and diversity, for the environment, for the resources of the planet we inhabit' (UNESCO 2005).

Editor's note: This is further expanded in Chapter 3.

Africa

South Africa—Morris Chauke

Sustainable development, according to the United Nations, means 'meeting the needs of the present generation without compromising the ability of future generations to meet their own needs' (World Commission on Environment and Development 1987). The task of protecting the diversity of species, genetic structures and ecosystems is based on three major factors:

1 The viability of ecosystems upon which current life forms and production processes are dependent requires sufficient biological diversity.
2 The needs of future generations are unknown.
3 Species required for critical processes in the future may now be unknowingly and wastefully driven to extinction (Munasinghe 1993).

Editor's note: Due to lack of resources, Morris Chauke was unable to fully complete his project, so his chapter is not included in this book.

Chapter summary

This chapter contains an overview or the project goals that form the basis of the nine country case studies reported in this book. Each of the authors is a scientist committed to sustainable lifestyles at global levels, and to helping their colleagues become more effective teachers. The collective wisdom of this group is introduced and the project background is outlined. In many ways this book reports the inspirational outcomes of a group of committed human beings determined to overcome all obstacles to bring learning and understanding to their respective communities. Many of the complexities facing the project are outlined in this chapter, along with the solutions arrived at during an intensive training period in the summer of 2005 in Rome. In the next chapter, an effort is made to scope out the philosophical dimensions associated with behavioural change in the whole of society, including issues related to agency and governance.

Bibliography

Beruchashvili, N, Chauke, M & Sánchez-Crispín, Á 2004, *Geographical perspectives on sustainable development*, International Geographical Union, Rome/Moscow/Beijing.

CEE (Centre for Environment Education) 2005, *The Ahmedabad Declaration*, CEE, New Delhi, viewed 5 June 2007, (http://www.ceeindia.org/esf/abad_d.asp).

LEAD 2007, *About LEAD*, LEAD, London, viewed 27 June 2007, (http://www.lead.org/page/5).

Munasinghe, M 1993, *Environmental economics and sustainable development*, World Bank, Washington DC.

Pile, S & Thrift, N (eds) 1995, *Mapping the subject: Geographies of cultural transformation*, Routledge, London.

Purvis, M & Grainger, A 2004, *Exploring sustainable development: Geographical perspectives*, Earthscan, London.

Robertson, M, Webb, I & Fluck, A 2007, *Seven steps to ICT integration*, ACER Press, Melbourne.

Roseland, M 2005, *Toward sustainable communities*, New Society Publishers, Gabriola Island.

UNESCO 2005, *Education for sustainable development: United Nations Decade (2005–2014): International implementation scheme (IIS)*, UNESCO, Geneva, viewed 22 May 2007, (www.unesco.org/education/desd).

Wilbanks, TJ 1994, '"Sustainable development" in geographic perspective', *Annals of the Association of American Geographers*, vol. 84, no. 4, pp. 541–56.

World Commission on Environment and Development (Brundtland Commission) 1987, *Our common future*, Norton, New York.

Websites

International Institute for Sustainable Development, (http://www.iisd.org)

UNESCO, (http://www.unesco.org)

United Nations, (http://www.un.org)

United Nations Division for Sustainable Development: Commission for Sustainable Development, (http://www.un.org/esa/sustdev/csd/policy.htm)

World Bank, (http://www.worldbank.org)

Notes

1   See the Foreword for details of the connection.
2   This funding was available through the International Council for Science (ICSU): see (http://www.icsu. org).
3   LEAD was established in 1991 by the Rockefeller Foundation. LEAD's international network claims nearly 1600 leaders from some 80 countries, and aims to work with them 'to mobilise others to make a real difference to the future of this planet' (LEAD 2007).
4   The Home of Geography at the Villa Celimontana in Rome is the international base for the International Geographical Union: see (http://www.homeofgeography.org).

2 | The gap between global rhetoric and local realities: How to build sustainable futures

Margaret Robertson

True education must correspond to the surrounding circumstances or it is not a healthy growth.

—Mahatma Gandhi

I have cherished the ideal of a democratic and free society in which all persons live together in harmony and with equal opportunities.

—Nelson Mandela

Mahatma Gandhi and Nelson Mandela are regarded as two of the most influential thinkers of the twentieth century. Now, as we move towards the end of the first decade in the twenty-first century, it is timely to reflect on their wisdom and the progress made within the global village of planet earth. The report card suggests that sustainable futures are moving further away from our community and that humans are largely to blame. Global warming, inequitable wealth, human degradation, poverty and disease, and the rapid growth of emerging economic power in the eastern cultures all contribute to a world that seems remote from philosophical rhetoric and in need of practical solutions that can be enacted now. What has happened to the political vision of our great leaders who struggled so selflessly during the last century to build a better, more sustainable future? How do we get back on course? Who can provide the leadership for this process in the twenty-first century?

The deeds and writings of Gandhi and Mandela provide reminders of what lies at the heart of humanity, regardless of time or place. Both men shared experiences of British colonialism and white power mixed with the local cultures of non-white indigenous populations. Both were scholars. They used their knowledge of theories and philosophy to embed their ideas. Illustrative is Mandela's perception of Marxist theory:

Today I am attracted by the idea of a classless society, an attraction which springs in part from Marxist reading and, in part, from my admiration of the structure and organization of early African societies in this country. The land, then the main means of production, belonged to the tribe. There were no rich or poor and there was no exploitation. (Mandela 1964)

In their lifelong commitment to confront the injustices of power and give a voice to all people regardless of colour, background, education or wealth, Mandela and Gandhi provided the world with new understandings of the democratic process. India, with its population in excess of one billion people, represents the 'biggest democracy' in the world. The majority black South African population now has the majority in the ruling African National Congress (ANC). Notwithstanding the acknowledged difficulties that remain in both countries, the outcomes represent the legacy of their commitment, shared ideals and unfailing sense of purpose to the healing process aimed at better and sustainable lifestyles.

In our more recent circumstances we can learn from their experiences. Researching their leadership qualities is instructive; several key qualities seem to be essential for leading local communities to overcome hardship and achieve better access to the basic essentials that sustain life—clean water, clean air and adequate food. These are:

- clear thinking about beliefs and their origins
- understanding the sovereignty of power and social structures
- knowing the economic and political spaces
- negotiating the goals
- engaging interest
- energising the process
- reflection.

Exploring these qualities will introduce the theoretical basis and reflective thinking required to successfully implement the case studies described in subsequent chapters.

1. The problem

The telltale global warming indicators point to unsustainable life habits (IPCC 2007a). There is need of urgent responses. Rapidly increasing development in the most populous nations of China and India and their associated pull on non-renewable energies, along with the excesses of existing wealthy nations, underlines the need for some hard decisions about how we live our daily lives. Water shortages in the world's biggest cities have spearheaded innovative practices related to water harvesting and recycling. However, the new leadership in this regard seems 'reactive'—reacting to need—rather than 'proactive'—developing practices to encourage less wastefulness. In developing nations, cities that are rapidly expanding through rural–urban migration have much more fundamental water issues related to infrastructure and the distribution of water for sanitation, drinking and basic health support. In any location, short-term needs can provide short-term answers. However, history books provide endless examples of corrupt powers applying bandaid short-term solutions. The wisdom of our great leaders indicates that long-term change takes time and may not occur in one generation.

Planned environmental management through education is at the heart of our book. Our holistic approach seeks responses that engage local people in projects that are meaningful on their terms, and mindful of their needs and skills. Having commenced the project under the unifying umbrella of the discipline of geography, we accept the need for working collaboratively with colleagues from all disciplines. Preparing the future for our planet requires the capacity of all thinking people. As Macnaghten and Urry (1998) observe, sustainability is the new public discourse. Indeed, the different perspectives of different people enrich the process of understanding this immense cause. Harvard Professor of English, Lawrence Buell, illustrates: 'Like racism, environmental crisis is a broadly cultural issue, not the property of a single discipline' (Buell 2005, p. vi). Adopting a phenomenological approach to 'subjective place-attachment' (p. 72), Buell is able to reveal multiple levels of mental mapping. There is *spatial orientation*, which brings in daily-lived space, and there is *temporal space* which we carry within through lived experience. This does not preclude the particular strengths of disciplines; indeed, Buell qualifies his view with an analysis of the particularities of disciplines which can help to solve various aspects of the puzzles. Within the discipline of geography, for example, Yi-Fu Tuan has long provided a role model for lateral thinking about environmental meaning-making (Tuan 1974). The aesthetics of the place or that which we may call 'home' can be interpreted from art and literature and a range of cultural artefacts.

The common advice from environmental thinkers is that there are critical ingredients for behaviours to change. These are:

- the influence of education and guided reflection on our personal subjectivities
- the need to plan, share and work together at local, national and global levels
- the constant and ongoing need for humility (reality checks)
- using the head and heart
- 'living the action'—a systems approach.

2. The power of education

Participation in mainstream society, regardless of location, requires knowledge, fundamental skills of literacy and communication, and a set of values that 'fit' the local, regional and national context. Developing nations struggle to spread their meagre resources. Within poor nations, the politics of budgetary decisions can make it very difficult for marginalised people to gain access to education, resources and infrastructure—such as fresh water, roads and communications—that can bring ideas and know-how to local people. Where education policy is ideologically driven, however, the evidence is that real change in people's capacity does take place.

Looking at the percentage of GNP spent on education, in Asia it is China, Thailand, India and Singapore which demonstrate high-level commitment to education. In Europe, Estonia demonstrates a similar commitment. Base-level data on percentage annual rise in GDP suggests these countries' policies are working. The old economies of Europe and North America, in contrast, appear to be slowing. While this is a somewhat simplistic overview that disguises the realities within each country of inequitable access to education, resources and monetary wealth, the trends indicate mobilising strength within the most populous nations of the world. Increased capacity to compete in the global arena of politics and economics, and resultant sharing of responsibility for the earth we all occupy, enables more people to be involved in local projects that can help remedy existing practices and provide shining examples for developing futures in emerging settings.

But considerable caution is needed before accepting the rhetoric of politics and the comfort that social indicators at national levels may bring. Grassroots realities cannot be forgotten. As Nisbett observed, 'If people do really differ profoundly in their systems of thought—their worldviews and cognitive processes—then differences in people's attitudes and beliefs ... might not be a matter merely of different inputs and teachings, but rather a consequence of using different tools to understand the world' (2003, p. xvii). This may be stating the obvious, but it may be the most sensible starting point for constructing dialogue that can legitimately and sustainably link local practices with global needs and principles. The projects described in this book used an acknowledgment of local expertise and views, and a readiness to understand, as fundamental starting points for educating communities. Education in this sense is more about building bridges in understanding. Language skills help the process of communication; communication can enable the learner and teacher to find common ground; this helps them construct the bridge for meaning-making; and this bridge can lead to improved knowledge. The following section illustrates how the world's educational communities are building capacity in education through such partnerships.

3. The need to plan, share and work together

We live in a world where global transactions affect all of humanity (OECD, 2005). Collaboration for sustainable futures is of crucial importance. Acceptance of the education imperative is the starting point; champions or leaders who can guide the people are also needed. Educating leaders for the challenge of bridge-building with local villages and communities in diverse spaces and locations across our vast planet is almost too much to comprehend. Such pessimism, however, does not help heal the planet. The starting point for the projects described in our book is the optimistic view that shared ground *does* exist—we just need to find it and build on its foundation.

One way to begin conceptualising is to examine the notion of 'territory'. People largely live their lives in bounded spaces. As Delaney (2005) comments, 'Territories are human social creations' (p. 10). Hence issues of sovereignty—both personal and public—take centre-stage when we consider society and nature. Delaney defines territory as a 'bounded meaningful space' (p. 15). The 'territory' may be a political entity, such as a country, or it could be a linguistic 'family'. The territory exists because it is 'meaningful' or, as Delaney describes it, these territories 'are significant insofar as they signify' (p. 14). 'Human territory', then, is about personal space and the meaning-making that comes from occupying a territorial position for a long time. From an educational perspective, understanding this concept is important. Personal meanings can assume the status of 'objective' facts, while personal 'subjectivities', or meanings, about the definition and role of 'knowledge' can get in the way of rethinking the ways in which we live our daily lives.

In the scientific literature, acceptance of contested space—including our definition of knowledge—has garnered considerable interest. Macnaghten and Urry (1998) provide a valuable reference for understanding the journey from localised views of nature and society to the present position where scientists, intergovernmental agencies and educators are generally agreed on the need to work together to fulfil their custodial responsibilities towards the planet and future generations of life on earth. Their research emphasises the 'ongoing debates on "risk", "trust" and agency' (Macnaghten & Urry 1998, p. 213). Changing behaviours comes back to personal choices about how we live our lives and a willingness to question the conventional wisdom. Modernist capitalist thinking (e.g. Harvey) has diminishing traction in the global politics of the twenty-first century. The educated middle class has entered the middle ground of politics with significant influence. Enter the strength of green politics, for example. The 'green' revolution has been gathering momentum for some years, effectively influencing outcomes within homes and parliaments around the world. Green politicians have lobbied governments and effectively influenced corporate behaviours to limit their fuel emissions. Effective media campaigns and protests have helped ensure the vibrancy of the agenda. Green politics is now embedded in the social fabric of education and daily living. When there is no defensible counterargument, it seems the politics of 'shame or blame' at local levels is influential. Habits related to the types of domestic products we purchase and dispose of through our home recycling operations are modelling lifestyle behaviours for young citizens. These messages are reinforced at school and legislated into workplace employment practices. In essence, we seem to have legislated social responsibility that brings global realities into the daily decision-making of individuals in local neighbourhoods. The two-way process of democratic decision-making reinforces the constructs of partnerships and linkages. This evolving dialogue of cooperation can be seen in the contributions of many intergovernmental agencies.

- **Organisation for Economic Co-operation and Development (OECD):** The publications of the OECD highlight the emergent strengths that can come from partnerships. Its *Civil Society and the OECD* policy statement declares: 'OECD ministers take the views that the OECD's role is to *enable* globalisation, make it inclusive and sustainable, and to seize the opportunities of open markets while at the same time addressing the needs of those who risk being left behind' (OECD 2005, p. 1, original emphasis). The rhetoric is about partnerships and cooperation from local to global and vice versa. Issues of trust, governance and agency permeate the documentation. The Internet is viewed as a significant resource for accessing the opinion of civil society as well as disseminating information. Member nations benefit from the cooperation, and from being aware of events and developments in other contexts. In educational circles the testing of students in participating nations provides ongoing advice on young people's attainment. The OECD Programme for International Student Assessment (PISA) collects data from member countries on students' knowledge, skills and competencies in reading, mathematics and science (OECD 2007).

- **Association of South East Asian Nations (ASEAN):** ASEAN is working in similar ways to promote cooperation and partnerships within its region. At the first meeting of Ministers of Education in the region in 1977, an educational network established a number of cooperative projects. The most recent meeting of the ASEAN Concord affirmed this stance with statements such as: 'The ASEAN Socio-cultural Community, in consonance with the goal set by ASEAN Vision 2020, envisages a Southeast Asia bonded together in partnership as a community of caring societies' (ASEAN 2003).

- **Asociación Latinoamericana de Integración (ALADI):** ALADI is the largest similar organisation in Latin America, with 12 member countries. Its articles include a commitment to cooperation on 'agricultural trade; financial, fiscal, customs and health cooperation; environment preservation; scientific and technological cooperation, tourism promotion; technical standards and many other fields' (ALADI n.d.).

- **Indigenous Peoples of Africa Co-ordinating Committee (IPACC):** In Africa, this is the network that brings the five major regional groupings of indigenous people together. Essentially a non-political organisation, IPACC is the collective 'voice' of indigenous Africans. Its stated aims further highlight the need for organisational structures that embed their principles in cooperation at local levels. These aims are to:

- *Promote recognition of and respect for indigenous peoples in Africa;*
- *Promote participation of indigenous African peoples in United Nations' events and other international forums;*
- *Strengthen leadership and organisational capacity of indigenous civil society in Africa in particular strengthening subregional networks of indigenous peoples ...*

The Bujumbura 2007 meeting of the IPACC Executive Committee also adopted a framework to strengthen the participation of indigenous peoples in environmental policy, protection and sustainable management and use of natural resources. (IPACC 2007)

- **United Nations Educational, Scientific and Cultural Organisation (UNESCO):** UNESCO aims to coordinate sustainable education programs throughout the world. Through its many agencies and programs, UNESCO promotes its vision that 'education is key to social and economic development. We work for a sustainable world with just societies that value knowledge, promote a culture of peace, celebrate diversity and defend human rights, achieved by providing education for all' (UNESCO 2007).
- **Trends in International Mathematics and Science Study (TIMSS):** In educational contexts, global monitoring and comparisons have long been occurring. TIMSS is conducted by the International Association for the Evaluation of Educational Achievement (IEA) and provides some baseline data on 60 nations. Many are from the developed world, including North America, Europe, Japan, Australia and New Zealand. But many participating nations are from developing regions including South-East Asia and the Middle East. The data provide benchmarks for progress and a forum for exchange of ideas related to curriculum and curriculum reform needs (IEA 2007).

In summary, there is ample evidence of the will to bring people together through shared visions of capacity-building education programs. One could conclude from this that the agency is in place. As illustrated in Chapter 1, there are multitudinous education projects running successfully all over the world. They demonstrate the power of partnerships for developing ways forward. For capacity-building projects aimed at sustainable development to be successful in developing nations, connection with local links is vital. All the case studies represented in this book reveal that local connections and partnerships were the key to their success.

4. The constant and ongoing need for humility

Developing good rhetoric is one thing, but having appropriate structures in place to provide the agency for activity is also important. However, very little happens if we are armchair leaders. Consider this scenario: a government of a wealthy nation has a policy of full employment and access to education at the highest levels for all its citizens. The land area is vast; there are many remote communities with limited communications access. Broadband digital communication helps bring the city services to the communities. The computers are boxed up and sent to the locals. All appears to be sorted—but no one thought about the need to show the local people how to turn the computers on! This is a simplified scenario, perhaps, but a stark reminder of the assumptions we can make in our decision-making.

At the 2002 international congress of the International Geographical Union, in his opening address Nelson Mandela enthralled participants with an account of a similar mistake. After his 27-year period of incarceration for political activism against apartheid he returned to his homeland, South Africa. The long period of imprisonment had insulated him from many of the issues he fought so courageously all his life. Observing a village woman filling her urn with water from the local waterhole, he recalled noting evidence of bacteria, possibly sewage:

> I said, 'But the water has tadpoles moving around, it has algae, this green stuff that covers stagnant water' … And then I asked this question: 'What do you do with this water before you use it?'
>
> They say, 'We do nothing. We use it as it is'.
>
> And then I asked a foolish question … 'Don't you boil it before you use it?'
>
> They all exclaimed simultaneously. 'Boil it with what? Look up, right up to the horizon, there is not a single tree. We have no electricity. With what must we boil it?' (Mandela 2002)

The simple assumption that energy sources were available as a normal part of daily life had become part of the accepted thinking of this long-serving prisoner. As unbelievable as it may seem, prison life in some ways was an improvement on local life. Electricity and clean water were accepted provisions.

Mandela's simple story has a powerful message. His is a lesson on deep humility, which serves as inspiration for all who seek to bring about change. Practical outcomes are not achieved or sustained through armchair politics and idealism. 'Feeling the pain' is about experiencing first-hand the reality of daily living and regularly revisiting the context. In brief, to achieve the plan requires understanding about the nature of meaning-making. Stephen Trudgill's (2003) analysis of 'meaning, knowledge and constructs in physical geography' identifies a number of dimensions of meaning-making that reinforce the collaborative nature of learning. 'Given meaning' is what we already know (p. 31) but 'discovery' is about 'finding a new meaning'. Hence, the 'experiential' component is a crucial part of this discovery. In scientific terms, this may come from fieldwork or structured observation of the spaces around us and how people behave within their spaces.

Authenticity is acknowledged as a fundamental requirement for educational success (Dewey 1916; LeFebvre 1991). Knowing the perspectives of our learners can often involve painful realisation of failed perspectives. For example, eco-feminist Vandana Shiva (1989) observed that '"development" was to have been a post-colonial project' (p. 253) or a 'new project of western patriarchy'. This view considers that 'Natural forests remain unproductive till they are developed into monoculture plantations of commercial species' (p. 255). There are many problems with this perspective, Shiva argues, not least that it further marginalises women:

> From the perspective of Third World women, productivity is a measure of producing life and sustenance; that this kind of productivity has been rendered invisible does not reduce its centrality to survival—it merely reflects the domination of modern patriarchal economic categories which see only profits, not life. (Shiva 1989, p. 255)

To bring about enhanced capacity is not about capital wealth using the western postcolonial paradigm. The pedagogy is wrong. As Freire (1972) maintains, pedagogy for the oppressed peoples of the world does not mean enforcing the dominant hegemony. To shift the oppressed towards a better way of life requires *dialogue*. Sustainable futures in this sense require letting go of current assumptions from both perspectives—rich and poor—then building shared meanings.

5. Using the head and heart

Once we acknowledge the power of 'other', the process of collaborating for better conditions should stand a better chance of success. Our case studies work from this assumption. The steps for successful project outcomes seem very straightforward. As Figure 2.1 identifies, all stakeholders need to keep in regular contact. Maintaining open channels of communication ensures that issues which arise during any of the steps of goal-setting, planning, implementation and evaluation of outcomes can be remedied. Each step requires dialogue, not assumptions.

The approach shown in Figure 2.1 acknowledges that wisdom is not the province of any one party. Wisdom may exist in having superior knowledge and skills about how to mobilise resources and develop infrastructure. It also exists in the hearts and minds of local people. Reluctance to let go of old practices, customs and habits needs to be viewed through construction of understanding. Working with the traditions of local people seems to be the commonsense approach to seeking solutions to problems which all parties agree need to be fixed.

Returning again to western thought processes, a small deviation into critical constructs of place-meaning and environmental management that have been influential is instructive. In the wake of World War II, Martin Heidegger's critique of the European narrative of modernity, without offering a 'next step', confronted the old–new paradigm. Heidegger defines place as 'dwelling' or 'living authentically' and as such can only be interpreted in situ (Hay 2002; Harvey 1996). The existentialism of being 'free' or independent of place does not fit the individual psyche (Neville

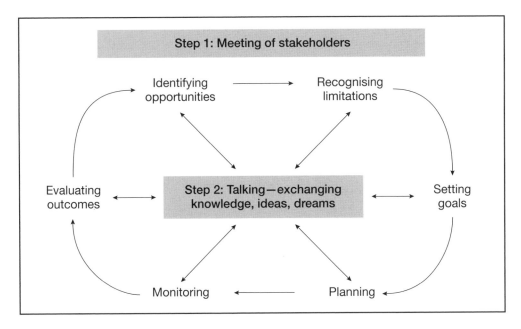

Figure 2.1 Steps for successful project outcomes

2005). This is vividly portrayed in a series of essays by individuals who have travelled the world since early childhood, titled *Unrooted childhoods: Memoirs of growing up global* (Eidse & Sichel, 2004). For many of the writers, the place of 'dwelling' is described in terms of favourite sneakers or other forms of portable attachment which could be adapted to each new location. 'Dwelling', in this nomadic lifestyle, educates the individual to expect differing customs, and different texts, through the intersection of language and culture. Returning to Heidegger's view, Harvey summarises this: 'Cultural life is then viewed as a series of texts intersecting with other texts, producing more texts' (1999, p. 312). In our places of dwelling, we are mediators of new texts and old thinking.

Environmental management, then, is as much about knowing *how* as it is about living the experience. Keen, Brown and Dyball (2005) are less abstract in their assessment. They remind us of the old adage that whole community involvement is needed 'to raise a child' (p. 3). Similarly, in the spirit of this belief, 'all members of society are needed to nurture a healthy environment' (p. 4). We can conclude from this brief overview that to achieve success in any project, the heart and the head need to work together.

6. 'Living the action'—systems thinking

The biodiversity crisis is not fiction. Loss of whole species and the declining numbers of many others are life-threatening conditions. The links between climate change and greenhouse gases are not fiction either (see the website references at the end of this chapter), nor are desertification and changing tidal patterns. For more than two decades we have been reminded of the impact of human activities on the planetary systems. The facts are striking. Schücking and Anderson, writing in the early 1990s, claimed that 'one fifth of all plant species on land face extinction within the next 20 years' (1991, p. 17). Many of the ecosystems involved play a critical role in the life systems (Buttimer, 2001; Pittock, 2005). Systems affected include coral reefs, forestry, water cycles, food chains and agriculture.

The Intergovernmental Panel on Climate Change (IPCC), the highly respected international forum for scientists from all related fields released its fourth assessment of climate change science (IPCC 2007b). Considered a somewhat conservative voice, the panel nonetheless concludes that human beings are changing the climate. Scientists now accept the need to collaborate to monitor the changes. While this chapter cannot do justice to the massive body of available literature, relevant website resources are listed at the end of this chapter. Most of these websites maintain active updates and are worth book marking for ongoing reference.

How do we respond? Writing on climate change, Pittock (2005) stresses the need to 'enhance adaptive capacity' as well as sustainable development. This view adds a practical dimension to the 'head and heart' construct. An adapted extract from his analysis (see Table 2.1) provides a practical approach to dealing with causes and consequences.

Table 2.1   Adaptive capacities of nature to climate change

Sector	Potential impact	Potential adaptation	Comment
Water resources	Increased drought and floods, changed water cycle patterns	Management, recycling, desalination plants	Climate change may shift the pattern
Coastal ecosystems	Salinity of coastal wetlands, catchment and biodiversity loss, invasion of exotic plants	Management of inputs, coral protection, effluent treatment	Growth in coastal population areas adds more environmental stress
Horticulture	Seasonal ripening patterns disturbed, new diseases and pests	Management including relocation, pesticides, increased efficiencies, changing enterprises	Climate change may cause food shortages. Carbon emissions may destroy the stock

Source: Adapted from Pittock (2005, p. 148)

From the table, it is clear that *management* is the common variable for enhanced adaptive capacity. Left alone, nature will find its own adjustment to global warming and related environmental responses at local levels. The impact from these activities is likely to be fall-out to the atmosphere and natural transport systems of the biosphere. This instability in nature points to the importance of understanding systems, and of using systems approaches in our actions on all levels of government and agency.

Consequently, there seems to be three concepts that need to be part of the thinking about adaptive change: *governance*, *activity* and *systems*.

- **Governance:** In the context of our interest in developing nations, an interesting collection of essays edited by Pugh (2000) shows how this pattern is emerging in developing countries where continuing growth of large cities has largely unchecked momentum. Building sustainable cities in the context of extreme poverty is a far cry from planning for whole-community change in rich nations. Pugh argues that the vision required is different. It needs to take into account the gaps between the economic forces of the city and the basic issues of poverty faced daily by ever-increasing numbers of new arrivals into the cities. Central to contemporary views on development studies is *governance*, which Pugh defines as 'primarily about steering policy and practice for improved social opportunity, welfare and economic efficiency' (pp. 8–9). It sounds simple when reduced to a few words. However, the policies needed, and the organisational, economic, educational and whole-of-society reforms required, are unimaginably complex. The need to start somewhere is obvious. Our case studies of water harvesting in Mumbai and water management in Beijing illustrate the contextually driven differences and need for locally designed responses.

- **Activity Theory:** One of the recognised holistic starting points for interpreting complexity is *Activity Theory*. Using the theoretical perspectives of the Russian psychologists Vygotsky and colleagues, Yrjö Engeström (2005) has developed the theory to apply within contemporary workplace, learning and organisational structures. The Center for Activity Theory and Developmental Work Research at the University of Helsinki views activity as the fundamental starting point for understanding functionally sustainable human systems. Activity Theory recognises the role of *motive* as the mediating element in action: 'People are not reduced to "nodes" or "agents" in a system; "information processing" is not seen as something to be modelled in the same way for people and machines'. Collective activity needs to be analysed by a series of actions, each driven by different motives. Hence adaptive capacity may require some humble starting points before whole-of-society change can be imagined, let alone attained. Of course individual actions can and do change, and indeed can become automated where systems are imposed collectively. Collective behaviour, such as driving a car, is a simple example of this. A contrasting example which illustrates the motives of individuals is the tobacco industry. Reduced rates of smoking in developed countries appear to be a successful outcome of health education and the cooperation of workplaces to meet healthy workplace guidelines. Threats of legal action help enforce smoke-free practices in public spaces. The response of the tobacco industry has been to shift its marketing strategy to developing nations, where the incidence of related dependency is growing.

- **Systems approach**: Taking the issue of governance into the reality of interpreting our complex life-sustaining systems is the great challenge. Building capacity requires accepting the social dynamic of a society. The application of Activity Theory can help start this process. *Systems approaches* seek to interpret relationships in space. Dyball, Beavis & Kaufman (2005) describe systems as having component parts: 'These might be physical things, such as animals, plants and rivers, or they might be conceptual, such as the various worldviews, attitudes, knowledge and beliefs held by different stakeholders' (p. 42). Furthermore, because both natural and built environments can be interpreted in terms of the flows and connections that link the parts, a 'systems' approach to analysing environmental matters is sensible.

   The other theoretical perspective that assists the process is *Complexity Theory*. As Axford reminds us, 'Global systems are networks of interaction that transcend both societal and national frames of reference' (2005, p. 188). Understanding the web of connections requires a vision that can accommodate this complexity. Complexity theorists argue that this vision needs to be incorporated into the role of agency and power relationships. The application of this approach in human interactions is recognition of the 'multidirectional movement of ideas' (Massey 2005, p. 126). Massey seems to agree with Thrift's analysis that 'complexity theory might well be being more successfully propagated outside natural sciences than within' (p. 127). In the end, the spatiality of interactions needs all the helpers it can muster to make processes and interactions meaningful in the domains of knowledge where relevant actions can occur.

7. Translating the 'grand plan' into action

This introductory chapter has considered the underlying principles of successful negotiation of action at local levels that can successfully integrate with global agendas for our fragile planet. The careers of great leaders have been instructive for starting this discussion. Mandela and Gandhi, along with other great leaders in history, provide inspiration—through their vision, their tenacity, their clarity of purpose, and their capacity to make complex dreams into simple truths. Great leaders mobilise oppressed people through encouraging them to perform self-sustaining actions. Analysing the speeches of such leaders reveals a common thread of argument—that individuals have the power to rise above hardship and survive.

   To halt the rapid assault on our living planet, enduring values and ideals of self-help need to be revived and redefined in today's local contexts, across the vast land

and seascapes of planet earth—and no part of the earth's surface is excluded from this challenge. Building humanitarian communities is an ideal we may all share. In pragmatic terms, the urgency is greater—our survival depends on the entire world's people building knowledge and understanding of the important responsibilities we all have to protect the planet for the future.

Bibliography

ALADI [n.d.], *About us: Overview*, ALADI, Montevideo, viewed 23 May 2007, (http://www.aladi.org).

ASEAN 2003, *Declaration of ASEAN Concord II (Bali Concord II)*, ASEAN, Jakarta, viewed 23 May 2007, (http://www.aseansec.org/15159.htm).

Axford, B 2005, 'Critical globalization studies and a network perspective on global civil society', in RP Appelbaum & WI Robinson (eds), *Critical globalization studies*, Routledge, New York, pp. 187–96.

Buell, L 2005, *The future of environmental criticism*, Blackwell, Oxford.

Buttimer, A (ed.) 2001, *Sustainable landscapes and lifeways*, Cork University Press, Cork.

Delaney, D 2005, *Territory*, Blackwell, Oxford.

Dewey, J 1916, *Democracy and education*, Macmillan, New York.

Dyball, R, Beavis, S & Kaufman, S 2005, 'Complex adaptive systems: Constructing mental models', in M Keen, VA Brown & R Dyball (eds), *Social learning in environmental management*, Earthscan, London.

Eidse, F & Sichel, N (eds) 2004, *Unrooted childhoods: Memoirs of growing up global*, Nicholas Brealey, London.

Engeström, Y 2005, *Developmental work research: Expanding activity theory in practice*, Lehmanns Media, Berlin.

Freire, P 1972, *Pedagogy of the oppressed*, Penguin, London.

Harvey, D 1996, *Justice, nature and the geography of difference*, Blackwell, Oxford.

—— 1999, 'Postmodernism', in A Elliott (ed.), *Contemporary social theory*, Blackwell, Oxford, pp. 303–17.

Hay, P 2002, *Main currents in western environmental thought*, UNSW Press, Sydney.

IEA 2007, *Trends in International Mathematics and Science Study 2007*, IEA, Amsterdam, viewed 23 May 2007, (http://www.iea.nl/timss2007.html).

IPACC 2007, *About IPACC*, IPACC, Cape Town, viewed 23 May 2007, (http://www.ipacc.org.za/eng/about.asp).

IPCC 2007a, *Intergovernmental Panel on Climate Change*, IPCC, Geneva.

—— 2007b, *Working Group III Fourth Assessment Report: Pre-copy edit version*, IPCC, Geneva, viewed 27 June 2007, (http://www.mnp.nl/ipcc/pages_media/AR4-chapters.html).

Keen, M, Brown, VA & Dyball, R (eds) 2005, *Social learning in environmental management*, Earthscan, London.

LeFebvre, H 1991, *The production of space*, trans. D Nicholson-Smith, Blackwell, Oxford.

Macnaghten, P & Urry, J 1998, *Contested natures*, Sage, London.

Mandela, N 1964, statement to Pretoria Supreme Court, Rivonia Trial, 20 April 1964, viewed 23 May 2007, (http://www.anc.org.za/ancdocs/history/rivonia.html).

—— 2002, address to International Geographical Union 2002, viewed 23 May 2007, (www.ucd.ie/geograph/AButtimer/MANDELAaward02.doc).

Massey, D 2005, *For space*, Sage, London.

Neville, B 2005, *Educating psyche: Emotion, imagination and the unconscious in learning*, Flat Chat Press, Greensborough, Vic.

Nisbett, RE 2003, *The geography of thought*, Nicholas Brealey, London.

OECD 2005, *Civil society and the OECD*, OECD, Paris, viewed 23 May 2007, (http://www.oecd.org/dataoecd/1/3/35744346.pdf).

—— 2007, *Programme for International Student Assessment (PISA)*, OECD, Paris, viewed 23 May 2007, (http://www.pisa.oecd.org).

Pittock, AB 2005, *Climate change: Turning up the heat*, CSIRO Publishing, Melbourne.

Pugh, C (ed.) 2000, *Sustainable cities in developing countries*, Earthscan, London.

Schücking, H & Anderson, P 1991, 'Voices unheard and unheeded', in V Shiva et al. (eds), *Biodiversity: Social and ecological perspectives*, Zed Books, London, pp. 13–41.

Shiva, V 1989, *Staying alive: Women, ecology and development*, Zed Books, London.

Trudgill, S 2003, 'Meaning, knowledge, constructs and fieldwork in physical geography', in S Trudgill & A Roy (eds), *Contemporary meanings in physical geography: From what to why?*, Arnold, London, pp. 25–45.

Tuan, YF 1974, *Topophilia: A study of environmental perceptions, attitudes and values*, Prentice-Hall, London.

UNESCO 2007, *Education strategy*, UNESCO, Geneva, viewed 23 May 2007, (http://portal.unesco.org/education/en).

Websites

Biodiversity

Biodiversity—Rainforest Portal, (http://www.rainforestweb.org/Rainforest_Information/Biodiversity)

Convention on Biological Diversity, (http://www.cbd.int)

Climate change

Climate Ark: Climate Change and Global Warming Portal, (http://www.climateark.org)

Intergovernmental Panel on Climate Change, (http://www.ipcc.ch)

International Decade for Action: Water for Life, 2005–2015, (http://www.un.org/waterforlifedecade)

United Nations Framework Convention on Climate Change, (http://unfccc.int/2860.php)

Greenhouse effect

Greenhouse Effect, Global Warming Facts and Greenhouse Gas Emissions Data Trends, (http://www.pewclimate.org/global-warming-basics/facts_and_figures/)

Desertification

United Nations Convention to Combat Desertification, (http://www.unccd.int)

Tidal change

Rising Tide UK: Factsheets, (http://risingtide.org.uk/resources/factsheets)

Data representation

For mapping using Geographical Information Systems, see ERIS: GIS and Mapping Software, (http://www.esri.com)

General

The World Conservation Union, (http://www.iucn.org)

3 | Education for sustainability: Defining the territory

Manuela Malheiro Ferreira

1. Defining sustainable development

The term 'sustainable development' was employed for the first time in 1972, on a report to the Club of Rome entitled *The limits to growth* written by a group of American scientists from MIT (Meadows et al. 1972). The group analysed the basic factors that could limit the growth on the planet: population, agricultural production, natural resources, industrial production and pollution. They concluded that:

> if the present growth trends in world population, industrialization, pollution, food production, and resource depletion continue unchanged, the limits to growth on this planet will be reached sometime within the next one hundred years. The most probable result will be a rather sudden and uncontrollable decline in both population and industrial capacity. (Meadows et al. 1972)

Basically, growing problems of resource depletion, pollution (including carbon dioxide concentration), loss of arable land and declining food production would converge to stop the progress of human society. However, they had affirmed that it would be 'possible to alter these growth trends and to establish a condition of ecological and economic stability that is sustainable far into the future' (Meadows et al. 1972). These scientists were the first to develop a model using computer technology and scientific method to analyse how growing human population and resource consumption were sustainable.

The ideas expressed in the *The limits to growth* were refused by other authors, who argued that the new ideas were alarmist—like the ideas of Thomas Malthus, published around 1800, that compared the linear increase in agricultural production with the geometric increase of population.

Twenty years later, the same group of scientists wrote a second book entitled *Beyond the limits* (1992). Considering additional evidence over the 20-year period, the authors affirmed that the world had entered a period of 'overshoot' in which it was well beyond sustainable levels of resource consumption, pollution and population—but a sustainable society was still technically and economically possible (Meadows, Meadows & Randers 1992).

The report of the World Commission on Environment and Development, commonly known as the Brundtland Commission, in 1987, brought worldwide attention to the need for promoting a development that does not endanger the environment and takes into account the existing resources. In the report, it is declared that:

> the overall aspirational goal must be sustainable development on the basis of prudent management of available global resources and environmental capacities, and the rehabilitation of the environment previously subjected to degradation and misuse. Development is sustainable when it meets the needs of the present without compromising the ability of future generations to meet theirs. (World Commission on Environment and Development 1987)

The Commission report created a strong basis for the United Nations Conference on Environment and Development held in Rio de Janeiro in 1992. However, as Wheeler and Beatley point out:

> [The Commission report] has been criticized on many grounds, particularly for accepting conventional notions of continued economic growth as the path to improve human welfare, for insufficiently incorporating an analysis of global power relations, and for developing a definition of sustainable development that is highly anthropocentric and dependent on the difficult-to-decide concept of 'needs'. (Wheeler & Beatley 2004)

The United Nations Conference on Environment and Development took place in Rio de Janeiro from 3 to 14 June 1992. Its goals were stated as:

> [E]stablishing a new and equitable global partnership through the creation of new levels of cooperation among States, key sectors of societies and people,
> Working towards international agreements which respect the interests of all and protect the integrity of the global environmental and developmental system,
> Recognizing the integral and interdependent nature of the Earth, our home. (UN 1992)

The Conference proclaimed 27 principles (see UN 1992), and an ample plan of action, the Agenda 21, was established. Agenda 21 is a plan to be taken on globally, nationally and locally by organisations in the United Nations system, by governments, and by major groups in every area in which humans impact on the environment (see UN 1993). The Earth Summit also led to the creation of a United

Nations Commission on Sustainable Development which meets annually to review international implementation efforts, and a United Nations Division for Sustainable Development to coordinate the agency's work in this field. The implementation of Agenda 21 and commitment to the Rio principles were reaffirmed at the World Summit on Sustainable Development (WSSD) held in Johannesburg, South Africa, in 2002. Some countries show political and financial commitments to implement Agenda 21; other national governments ignore its principles. Nevertheless, Agenda 21 promoted global debates on sustainability and the foundation for global action.

In 2002, the United Nations officially adopted the Millennium Declaration (UN 2000). Eight main goals and 18 related strategies for the global development agenda were defined. The goals are:

- *Goal 1: Eradicate extreme hunger and poverty*
- *Goal 2: Achieve universal primary education*
- *Goal 3: Promote gender equality and empower women*
- *Goal 4: Reduce child mortality*
- *Goal 5: Improve maternal health*
- *Goal 6: Combat HIV/AIDS, malaria and other diseases*
- *Goal 7: Ensure environmental sustainability*
- *Goal 8: Develop a global partnership for development. (UN 2006)*

One target of Goal 7 (ensure environmental sustainability) is to integrate the principles of sustainable development into country policies and programs and reverse the loss of environmental resources.

Despite all the agreed principles in international forums, the concept of 'sustainable development' has been used in different ways. Cabral (2002) identifies three uses:

1   As a theoretical concept, in terms of **structurally relating environment and development** problems through cause/effect and cyclical relationships. The problems delivered through the environment result in environmental degradation and knock-on problems for society in terms of a diminished resource base and further impoverishment and inequity.
2   As a **set of goals,** shared and conflicting (depending on the ideology of the proponents) about, for example, the amelioration of poverty, political democracy, and conservation and regeneration of the environment—all in the context of intragenerational and intergenerational equality.
3   As a **strategy,** or a process, for achieving the above goals.

Education for Sustainable Development (ESD) takes into account these three uses of 'sustainable development', because it includes an analysis of the concept of sustainable development and of its goals, and tries to enhance people's views of how to achieve the goals at different scales, from the local to the global.

2. Education for Sustainable Development (ESD)

The period from the 1 January 2005 to 31 December 2014 was declared by the United Nations as the Decade of Education for Sustainable Development (DESD). UNESCO is the lead agency of this Decade.

Education for sustainable development (ESD) has its roots in the history of two distinct areas of interest of the United Nations—education and sustainable development. The draft International Implementation Scheme (IIS) for the Decade defined its objectives:

- *facilitate networking, linkages, exchange and interaction among stakeholders in ESD;*
- *foster an increased quality of teaching and learning in education for sustainable development;*
- *help countries make progress towards and attain the millennium development goals through ESD efforts;*
- *provide countries with new opportunities to incorporate ESD into education reform efforts. (UNESCO 2005)*

The draft IIS also discusses the general aims of the Decade:

The overall goal of the DESD is to integrate the principles, values, and practices of sustainable development into all aspects of education and learning. This educational effort will encourage changes in behaviour that will create a more sustainable future in terms of environmental integrity, economic viability, and a just society for present and future generations ...

Education for Sustainable Development (ESD) prepares people of all walks of life to plan for, cope with, and find solutions for issues that threaten the sustainability of our planet ... Understanding and addressing these global issues of sustainability that affect individual nations and communities are at the heart of ESD. These issues come from the three spheres of sustainable development—environment, society and economy. (UNESCO 2005)

In relation to values, it is stated that:

The ways countries decide how to approach sustainable development will be closely linked to the values held in these societies, for it is these values that define how personal decisions are made and how national legislation is written. Understanding values is an essential part of understanding an individual's own worldview and that of other peoples. Understanding your own values, the values of the society you live in, and the values of others around the world is a central part of educating for a sustainable future. Each nation, cultural group, and individual must learn the skills of recognizing their own values and assessing these values in the context of sustainability.

United Nations history carries with it a host of values related to human dignity and rights, equity, and care for the environment. Sustainable development takes these values a step further and extends them between generations. With sustainable development comes valuing biodiversity and conservation along with human diversity, inclusivity, and participation. In the economic realm, some embrace sufficiency for all and others equity of economic opportunity. Which values to teach and learn in each ESD programme is a matter for discussion. The goal is to create a locally relevant and culturally appropriate values component to ESD that is informed by the principles and values inherent in sustainable development. (UNESCO 2005)

The priorities for ESD presented in IIS are improving access to quality basic education; reorienting existing education programmes; developing public understanding and awareness of sustainability; and providing training in all sectors of activity.

The key characteristics of ESD, according to the same document, are as follows.

No universal models of ESD exist. While there is overall agreement on principles of sustainability and supporting concepts, there will be nuanced differences according to local contexts, priorities, and approaches. Each country has to define its own sustainability and education priorities and actions. The goals, emphases and processes must, therefore, be locally defined to meet the local environmental, social and economic conditions in culturally appropriate ways. Education for sustainable development is equally relevant and important for both developed and developing countries.

ESD has essential characteristics that can be implemented in many culturally appropriate forms.

Education for sustainable development:

- *is based on the principles and values that underlie sustainable development;*
- *deals with the well being of all three realms of sustainability—environment, society and economy;*
- *promotes life-long learning;*
- *is locally relevant and culturally appropriate;*
- *is based on local needs, perceptions and conditions, but acknowledges that fulfilling local needs often has international effects and consequences;*
- *engages formal, non-formal and informal education;*
- *accommodates the evolving nature of the concept of sustainability;*
- *addresses content, taking into account context, global issues and local priorities;*
- *builds civil capacity for community-based decision-making, social tolerance, environmental stewardship, adaptable workforce and quality of life;*
- *is interdisciplinary. No one discipline can claim ESD for its own, but all disciplines can contribute to ESD;*
- *uses a variety of pedagogical techniques that promote participatory learning and higher-order thinking skills.*

These essential characteristics of ESD can be implemented in myriad ways, so that the resulting ESD programme reflects the unique environmental, social and economic conditions of each locality. Furthermore, ESD increases civil capacity by enhancing and improving the workforce, social tolerance, environmental stewardship, participation in community-based decision-making, and quality of life. To increase civil capacity in these five areas, formal, non-formal and informal education must be combined. (UNESCO 2005)

ESD concerns every country and everyone. Therefore its implementation needs to take into account that the concept of 'sustainable development' defined by the Brundtland Commission (development that 'meets the needs of the present without compromising the ability of future generations to meet theirs') is not universal but varies between different communities. It depends on each community's economic, social and environmental realities, and their values and attitudes which are linked to their cultural characteristics. In this way, problems of development and of sustainability can only be solved through the active participation of informed citizens who look at the realities at different scales—local, regional, national and even global. These are citizens who understand not only the relationship between environmental, economic and social aspects, but also between the ways people live in their community and in other communities. Citizens have to take into account both the needs and rights of the present and future generations; understand the relations between power, resources and human rights; be able to evaluate the consequences, at different levels, of the diverse styles of life of the populations; and be able to evaluate different answers that individuals and organisations give to different problems of local scope. ESD requires conceptual learning and the development of skills, values and attitudes. Local problems with regional, national and even global consequences should be addressed, and schools, local authorities and citizens should be involved in their resolution.

3. School learning

At school level, sustainable development issues should be studied at different levels and take into consideration different dimensions (personal, social, spatial, temporal, economic, political, historical, cultural and aesthetic). Students need to explore how environmental issues relate to their everyday lives and how lifestyles have an impact on environment. They should develop skills in critical thinking, problem-solving and decision making, and values in relation to the environment and to development. Teachers should use different strategies, such as cooperative learning, enquiry-based learning, problem-solving, fieldwork, role-playing and simulations. Students should be involved in real processes of decision-making and action particularly at the local level, but also at the regional, national or even global level (Ferreira 2004).

Geography can take an important role in ESD. Frances Slater points out that:

geography in education is both a cognitive and value laden affair through our very selection of what we study and on which we exercise our cognitive faculties ... Geography in education like all subjects is not neutral. Its substance, as one source of its value ladenness, does not stand apart from our constructions and interpretations of spatial and environmental relations. So geography is shot through with a value bias, linked among other things, to people's experience, perception and conception of their environment, how they evaluate it and seek to live in it. (Slater 2002)

The relation between cognitive and value aspects of ESD is central to the education of citizens who would like to live in a world where development meets the needs of the present generations without compromising the ability of future generations to fulfil their own needs.

4. Environmental education in Europe: Application of the agreed principles

Since 1974, UNESCO, in conjunction with the United Nations Environment Programme (UNEP), has had a key role in promoting environmental education. The official starting point for environmental education was the Conference of Belgrade, 1975, followed two years later by the intergovernmental conference of Tbilisi. UNESCO and the UNEP sponsored a series of pilot research studies and created the first documentation for teacher training in environmental education. To support and enrich their action, they also organised a series of regional conferences as well as another intergovernmental conference—this time in Moscow (1987)—which were undoubtedly very important in this institutional phase.

In Europe in the 1980s, the EEC and the Council of Europe sponsored a network of pilot schools. Unfortunately, these initiatives did not transfer to other schools, and were only undertaken by some teachers in some classes. At the same time, the mass media began alerting the public to a range of environmental issues such as pollution, nuclear waste and desertification, and to important global questions related to the biosphere, such as population growth, the depletion of the ozone layer and the greenhouse effect. However, the media's impact on widespread environmental education was limited. It became clear that while many governments, administrative bodies and private companies understood the importance of the ecological movement, and promoted ecological values in their documentation, the principles they declared often went no further than advertising.

Nevertheless, national and regional administrations played a more positive part by creating structures and places for environmental activities. These are located largely in the domain of education—that is, in schools. For example, a government authority would organise classes in conjunction with, say, an ecological museum, a scientific club, or a national or regional park education centre (Giordan n.d.). So within the teaching discipline, changes were made: curriculum and syllabus reformulations, including geography and biology; production of new environmental teaching materials; development of activities for students; and training of teachers.

The European Union (EU) has an important role in Europe's environmental policy, and environmental education is part of its effective implementation. A report was commissioned by the Environment Directorate-General of the European Commission, titled *Environmental education in the educational systems of the European Union*. Research was undertaken by Eleanor Stokes, Ann Edge and Anne West in 2000 and 2001. According to the researchers, during the two last decades:

> the concept of 'the environment' has changed over time. Early views focused on changing ecosystems and the impact of various forms of pollution. However, the social, economic and cultural dimensions of the environment have been increasingly recognised and the inclusion of sustainable development makes the concept even more broad. (Stokes, Edge & West 2001, p. 4)

According to the report, the study revealed a diversity of ways in which member states of the European Union address environmental education in the primary and secondary school systems. The main conclusions are:

> Environmental education is a compulsory area of the curriculum in primary and lower secondary education. It is taught using a range of different approaches, the most common being that it is embedded in other subject areas, in particular geography and the sciences (notably biology). In some countries an interdisciplinary thematic approach is used and in a number of cases this is combined with other approaches, notably the embedding of the subject material in other subject areas. (Stokes, Edge & West 2001, p. 27)

The authors also point out:

> It is interesting to note that in addition to broad areas of knowledge in relation to environmental education, the importance of values, ethics, attitudes and behaviours in the curriculum emerges, thus giving the teaching of environmental education a perspective not always found in other areas of the curriculum. This approach suggests that general concerns about the environment and sustainability are being taken seriously by policy makers striving to inculcate attitudes and values that will result in environmentally responsible behaviour by young citizens of Europe.
>
> At the upper secondary level, a different approach emerges. Here, the curriculum became increasingly differentiated. In a number of countries there are specialist courses for environmental studies. In addition, the subject material is embedded in various other broad areas of knowledge—in particular, geography and the sciences. ...
>
> [I]t is clear that there are points of convergence and divergence in relation to the teaching of environmental education across the European Union—particularly in the way in which it is embedded across the curriculum and in terms of the tendency for it to address values, ethics, attitudes and behaviour. There are some interesting points of divergence, particular in relation to whole school initiatives. These are clearly intended to encourage responsible behaviour by young people. (Stokes, Edge & West 2001, pp. 27–8)

Today, due to the fact that ten more countries have since joined the European Union, the situation is even more diversified. Associations of nature protection, ecologists and consumers have developed in Europe, and have, without any doubt, largely contributed to the growing awareness in the population of environmental issues and the need to develop environmental education. Among these associations is the Foundation for Environmental Education (FEE), formerly known as the Foundation for Environmental Education in Europe (FEEE), which was established in 1981. In the early years of FEEE, the organisation was mostly active through internal meetings, external seminars, conferences and a number of publications. This association developed campaigns, including the Blue Flag Campaign, the Eco-Schools Campaign, the Young Reporters for the Environment Campaign and the Learning about Forests Campaign. The last three were especially developed in school contexts.

The Blue Flag Campaign was originally a French idea, launched in 1985. In 1987, it was launched on a European level (10 countries) in a partnership between FEEE and the European Commission. According to FEE:

> The Blue Flag works towards sustainable development at beaches/marinas through publicly [rewarding] sites that meet strict criteria dealing with water quality, environmental education and information, environmental management, and safety and other services. A few of the issues covered in the criteria include cleanliness, provisions for waste and recycling, zoning of activities, and environmental education activities for a variety of people. (FEE n.d.)

In 1994, the Eco-Schools Campaign and the Young Reporters for the Environment Campaign were implemented as second and third campaigns of FEEE. Eco-Schools is a program which aims to help schools improve their immediate environment, and the environment of their local communities. Its seven-step process engages students and teachers in setting goals, monitoring environmental performance and evaluating outcomes. Its combination of learning and action has the potential to influence the lives of young people, school staff, families and local authorities.

The Young Reporters for the Environment (YRE) is a series of groups, established in secondary schools, stretching across 17 countries. Each group takes on an investigation project about an environmental issue, then communicates information about that issue to the public. YRE operates on two levels. At the local level, students investigate a problem and link environmental and scientific issues. They report their findings to the local community through the local newspapers, radio, television, conferences or public exhibitions of their work. The second level is international, where students meet young reporters from other countries, and use the Internet to communicate, share information and produce jointly written articles.

The Learning about Forests Campaign was launched by the FEEE in 2000. It aims to encourage school classes and teachers to use forests for educational activities—to visit forests, learn about them and in them, and share their experiences across the world.

In the mid-1990s there was a growing interest in FEEE programs outside Europe; in the 2000s, a number of non-European organisations joined FEEE. The organisation could no longer limit itself to Europe, so it dropped the last 'E' in its name and became the Foundation for Environmental Education (FEE). In 2003, FEE signed a Memorandum of Understanding with UNEP. This allows the two organisations to cooperate in 'areas of common interest relating to education, training and public awareness for sustainable development globally' (FEE n.d.).

5. What are the messages for environmental education?

According to André Giordan, today education for the environment is considered more and more frequently as necessary, but its principles have not yet become widespread in all schools. Giordan argues that to improve environmental education, it is not enough to simply organise more activities; the contents and objectives of education need to be re-examined, and curricula and syllabuses need to be reorientated to ensure a systemic approach (Giordan n.d.). It is clear, too, that teachers, journalists and organisers need to be trained in environmental issues, values and ethics. Environmental education needs to be assessed and evaluated, just as other school subjects are. As well as its environmental elements, education for sustainable development also includes social, economic and cultural aspects, and as such is firmly linked with the growing trend of citizenship education.

Bibliography

Cabral, S 2002, 'Sustainable development and education: what is it all about?', in *Proceedings of Socrates/Comenius course: Active citizenship, sustainable development and cultural diversity*, Universidade Aberta, Lisbon.

FEE [n.d.], *About Blue Flag*, FEE, Copenhagen, viewed 24 May 2007, (http://www.fee-international.org/Programmes/blueflag).

Ferreira, M 2004, 'Cultural diversity', in B Miranda, F Alexandre & M Ferreira (eds), *Sustainable development and intercultural sensitivity: New approaches for a better world*, Universidade Aberta, Lisbon.

Giordan, A [n.d.], *L' Education pour l'Environnement en Europe*, Laboratoire de Didactique et d'Epistémologie des Sciences (LDES), Geneva, viewed 24 May 2007, (http://www.ldes.unige.ch/publi/vulg/EEE.htm).

Meadows, DH, Meadows, DI & Randers, J 1992, *Beyond the limits*, Chelsea Green, Post Mills, VT.

Meadows, DH, Meadows, DI, Randers, J & Beherens, W III 1972, *The limits to growth*, Universe Books, New York.

Slater, F 2002, 'Citizenship education through geography: Values and values education in the geography curriculum in relation to concepts of citizenship', in *Proceedings of*

Socrates/Comenius course: Active citizenship, sustainable development and cultural diversity, Universidade Aberta, Lisbon.

Stokes, E, Edge, A & West, A 2001, *Environmental education in the educational systems of the European Union*, European Commission, Brussels, viewed 24 May 2007, ⟨http://ec.europa.eu/environment/youth/original/pdf/envedu_en.pdf⟩.

UN 1992, *Report of the United Nations Conference on Environment and Development: Annex 1: Rio Declaration on Environment and Development*, viewed 23 May 2007, ⟨http://www.un.org/documents/ga/conf151/aconf15126-1annex1.htm⟩.

—— 1993, *Agenda 21: Earth Summit—the United Nations programme of action from Rio*, UN, Geneva, viewed 23 May 2007, ⟨http://www.un.org/esa/sustdev/documents/agenda21⟩.

—— 2000, *United Nations Millennium Declaration*, UN, Geneva, viewed 23 May 2007, ⟨http://www.un.org/millennium/declaration/ares552e.pdf⟩.

—— 2006, *UN Millennium Project: About the MDGs*, UN, Geneva, viewed 27 June 2007, ⟨http://www.unmillenniumproject.org/goals/index.htm⟩.

UNESCO 2005, *Draft International Implementation Scheme (IIS) for the United Nations Decade of Education for Sustainable Development (2005–2014)*, UNESCO, Geneva, viewed 5 June 2007, ⟨http://unesdoc.unesco.org/images/0014/001403/140372e.pdf⟩.

Wheeler, SM & Beatley, T 2004, *The sustainable urban development reader*, Routledge, London.

World Commission on Environment and Development (The Brundtland Commission) 1987, *Our common future*, Norton, New York.

Websites

European Commission: Environment, ⟨http://ec.europa.eu/environment/index_en.htm⟩
Foundation for Environmental Education (FEE), ⟨http://www.fee-international.org⟩

4 | Sustainable development: What it means in the global context

WU Shaohong

1. Background and problems

In 1987 the World Commission on Environment and Development advocated 'sustainable development' in the report titled *Our common future*. Though there were many definitions of sustainable development after that, the core ideology of sustainable development is the 'development that meets the needs of the present without compromising the ability of future generations to meet their own needs' (World Commission on Environment and Development 1987). In the last two decades the ideology of sustainable development has been widely developed. The common recognition is that sustainable development is based on equality between generation and generation, human and nature, region and region, and sector and sector.

However, sustainable development is not fabricated. It appeared only after humankind realised the damage being done to resources and the environment. In prehistory, humans clung to nature. In the agricultural age, they started using and developing nature. This development was within the tolerances of nature because of a lower technological level. From the industrial age, especially since the Industrial Revolution, rapid advances in science and technology have enabled humankind to explore resources, develop nature and change the environment. Gradually, humans went too far in developing nature, believing that they could take and use nature without limitation, even believing that nature could be changed as well as conquered. Such ideology and behaviours have brought severe damage to the environment, and even threatened the existence and development of humans.

2. Biodiversity decrease

Humans cut and destroyed forests for fuel and building materials, cultivated and overgrazed grassland for food and hunted wildlife, which led to land desertification, salinisation and leanness, as well as ecosystem simplification. Such activities also affected the existence and development in space of living species, and led to the loss of species and generic resources. According to international organisations such as The World Conservation Union (IUCN), research on avifauna has shown that one genus of bird became extinct every 300 years from 35 million to one million years ago, every 50 years from one million years to modern times, every 2 years during the last 300 years, and every year during the twentieth century. There are 20 000 to 30 000 kinds of flowering plants and over 1000 kinds of animals in severe danger all over the world. By the end of the twenty-first century it is estimated that between 500 000 and one million species will become extinct. One species disappeared from the earth every day in the 1980s and every hour in the 1990s. Half of the species on earth will probably be destroyed by humans in the next 50 years (Liu & Li 2001).

3. Soil erosion

Human-induced soil erosion is mainly due to socioeconomic activities such as deforestation, grassland cultivation and overgrazing. The main consequences of soil erosion are water loss; frequent flood and drought; soil loss induced by aggravated ecocatastrophes; and soil nutrient loss induced by soil leanness, salinisation and the eutrophication of surface water. According to statistics of the United Nations Environment Programme (UNEP), worldwide soil loss by erosion is about 60 billion tonnes per year. If the average global soil thickness is, say, one metre, then all the soil will be lost in 809 years. About 40 per cent of the total soil erosion flows into the sea, and the other 60 per cent is deposited in rivers, lakes and reservoirs.

4. Land desertification

The most significant cause of land desertification is the damage to grasslands located in semiarid or arid regions, which are characterised by poor natural conditions and fragile ecosystems. During the last 50 years, seven million square kilometres of desert resulted from damage to grassland. The area of desert caused by human activities is equal to the total area of Brazil, and the annual increase is equal to the total area of Sri Lanka. There are 1.2 billion people—accounting for about 20 per cent of the total world population—who are threatened by desertification.

5. Forest reduction

Forest reduction and disappearance enhances the greenhouse effect, land degradation, soil erosion, biodiversity decrease (caused by habitat loss), and ecosystem imbalance. Despite these serious consequences, global forest areas are still disappearing. Tropical forest accounted for 1.935 billion hectares in 1980, but 45 per cent of it was severely damaged. According to statistics from the Food and Agriculture Organization (FAO), the vanishing tropical forest amounted to 11.3 million hectares each year at a decreasing rate of 0.58 per cent from 1981–85, and 16.78 million hectares each year at a decreasing rate of 1.2 per cent in 1990. From 1950–80, over half of the forests in the world were destroyed (Ma 1999; Liu & Li 2001). From 1990–2000, the globally averaged decreasing forest area was about 0.78 per cent (FAO), which was 1.1661 million square kilometres from a total of 149.5 million square kilometres.

6. Greenhouse effect

Since the Industrial Revolution, the main greenhouse gases in the atmosphere have obviously increased. Before the Industrial Revolution, carbon dioxide in the atmosphere was about 280 parts per million (ppm), and methane was 715 parts per billion (ppb). In 1958, carbon dioxide was 350 ppm but up to 379 ppm in 2005. In the early 1990s, methane was 1732 ppb and up to 1774 ppb in 2005. The global atmospheric nitrous oxide concentration increased from a pre-industrial value of about 270 ppb to 319 ppb in 2005. The concentration of industrial fluorinated hydrofluorocarbons (HFCs) is relatively small but increasing rapidly. Consequently, over the past 100 years (1906–2005), the global temperature has risen about 0.74°C. Based on current socioeconomic development, many scientists predict that by 2100 atmospheric carbon dioxide concentrations will be up to 730–1020 ppm for the standard SRES (Special Report on Emissions Scenarios) A2 scenario in the coupled climate-carbon cycle models. This means that, at the end of the 21st century, temperature will probably increase about 1.1–6.4°C (IPCC, 2007).

Greenhouse-induced consequences include abnormal climate, such as the El Niño and La Niña phenomena, frequent crop disease and pest infestations, declining farm productivity, and frequent natural disasters such as flood, drought and extreme temperature changes. Another serious consequence is ice and snow thawing quickly in the two polar areas and in high mountain areas such as the Himalayas, which will cause sea levels to rise and threaten countries, regions and cities along coastal areas. These include Holland, Bangladesh, Egypt, the Maldives, the Seychelles, New York, Rio de Janeiro, Venice, Alexandria, Shanghai, Hong Kong, Tokyo and Bangkok.

7. Ozone depletion

Ozone, a special composition of gas in the atmosphere, covers the earth to form a protective layer. Its important function is to absorb most ultraviolet rays from the sun to guarantee normal growth and continuation of species. Scientists have proven that the ozone layer in the atmosphere has been depleted 3 per cent each year by more than 30 substances, especially Freon, methane and carbon monoxide

(Ma 1999). The decreasing ozone attenuates the ozone layer, and some ozone holes have appeared. For example, in Antarctica in 1995, an ozone hole in the atmosphere lasted more than 40 days and covered an area of 2000 square kilometres. Another ozone hole with a diameter of 1000 kilometres appeared over southern Chile in 1997. When ozone is depleted, ultraviolet rays from the sun increase, raising the incidence of skin cancer, causing the mass death of fish in the sea, reducing or even cutting off crop production by damaging crop organisms, distorting and shortening the lifetime of some chemical and plastic products, and inducing greenhouse effects.

8. Sand-dust weather

Human activities sometimes play a more important role than natural factors in inducing sand-dust weather. Sand-dust storms often occur around the world because of loose substances blowing into the atmosphere after surface vegetation is destroyed in arid or semiarid regions. The famous 'black storm' in the USA in the 1930s was a typical example of a sand-dust storm. The frequently occurring sand-dust weather from central Asia has endangered eastern Asia, especially eastern China.

9. Frequent natural disasters

Frequent natural disasters, including earthquakes, flood, typhoons and storm tides cause great loss to the world every year. Irrational human activities and natural variation accelerate the occurrence of natural disasters; at the same time, they reduce natural defences to disaster and aggravate losses. The direct economic loss caused by natural disasters in the world in the last century is shown in Figure 4.1.

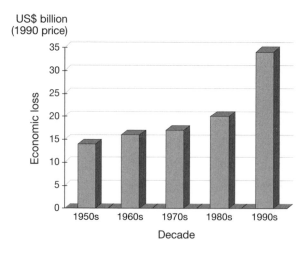

Figure 4.1 Economic loss caused by natural disasters in the last 50 years

Source: Ma (1999)

10. Air pollution

Several processes release exhaust emissions or noxious gases into the atmosphere, including industrial development, agricultural activities such as rice planting, the use of coal for fuel, vehicle emission, Freon production and some natural phenomena such as volcanic eruptions. The emitted gases mainly consist of sulphur dioxide, nitrogen oxide, carbon dioxide, carbon monoxide, methane, plumbum, hydrocarbon, soot and dust. In addition to the direct effect on human health through breathing, air pollution contributes to the greenhouse effect, ozone depletion and acid rain, thereby inducing new environmental problems. The annual emission of sulphur dioxide is 1500 million tonnes per year, which generates acid rain, acid snow, acid mist and fine aerosol particles causing severe surface water acidification.

11. Water pollution

Water pollution mainly comes from:

- the direct discharge of industrial effluents
- the eluviation of agrochemicals such as fertiliser
- pesticides and herbicides
- atmospheric substance deposition
- leakage from traffic, transport and industry
- the discharge of domestic sewage, garbage and solid waste.

Among these processes, industrial effluents are the most severe. Freshwater pollution is the most critical threat to humans. About 98 per cent of global water is salt water which cannot be utilised for agriculture, daily living and most industries. Most of the remaining scarce freshwater accumulates in the two polar regions, and underground. Therefore only 0.014 per cent of surface freshwater can be utilised directly and the pollution of it would threaten human survival. Marine pollution is another important part of water pollution. Eutrophication caused by too much organism discharge into the sea interferes with the ecosystem and even causes death to marine life. Other important causes of marine pollution are marine transport, marine oil extraction and marine mining. Throughout the world there are about 10 million tonnes of petroleum, 0.025 million tonnes of chlorinated biphenyls, over 3.9 million tonnes of zinc, over 0.3 million tonnes of lead and 0.25 million tonnes of copper released into the oceans each year.

12. Soil pollution

Soil pollution is mainly caused by chemical products used in agriculture, mineral solid waste, living garbage accumulation and industrial dust deposition. The pollutant run-off from soil erosion pollutes surface water. Most pesticides and herbicides have toxicity in different degrees, and are difficult to degrade so they remain in the soil for as long as 200 years. Large amounts of living garbage are accumulated on land and find their way into the soil. Gangue from mining, waste residues and solid wastes from industries are also dumped, and even waste water and dust from factories is discharged. All the above activities cause chemical, physical and biological pollution in the soil. Since World War II, humans have damaged about 10.5 per cent of the area of productive land, two-thirds of which has been severely destroyed, and 300 million hectares of land is no longer arable.

13. Noise and electromagnetic wave pollution

Noise comes from factories, engineering works, vehicles, trains and aircraft. When the frequency and amplitude of sound exceeds a certain range (e.g. too frequent or too loud), it makes people feel ill at ease (dysphoria), and may even lead to eardrum perforation, hearing loss and other symptoms. Communication equipment, transmission lines and industrial facilities form an electromagnetic radiation field covering the earth.

14. Energy crisis

Energy consumption is out of balance—for example, in 1992, the developing countries with 80 per cent of the world's population consumed 35 per cent of its primary energy. Traditional energy sources have dried up. Petroleum, fossil fuels and natural gas are the current dominant but non-renewable resources. The proven world petroleum reserves are 700 billion barrels and natural gas reserves are 5.5 billion stere. These could last 34–40 years at the present consumption rate. We may not exhaust the energy reserves after 40 years, but the resources will eventually run out. The use of petroleum and fossil fuels causes environmental pollution and contributes to the greenhouse effect. Therefore, at the 1992 Rio de Janeiro Conference, the United Nations Framework Convention on Climate Change regulated that the greenhouse gases in the atmosphere should be stabilised at 'a level that would prevent dangerous [human] interference with the climate system' (UNFCCC 1992, p. 4). According to the objective, the emission of greenhouse gas should decrease 60–80 per cent of the level in 1992. The decrease would greatly benefit energy consumption and economic profit.

15. Problems in mineral resources use

The exploitation and utilisation of mineral resources at present has many problems which threaten the environment, including:

- water and air pollution during mining and manufacturing processes, which endangers human health
- large areas of land being occupied
- soil pollution from strip mining
- rock mass structure damage, causing cracking and collapse, leading to frequent natural disasters in underground mining
- marine pollution rise because of discharge and leakage during marine mining and marine transportation.

16. Water resources shortage

As early as 1977, the United Nations warned the world that water would become a severe social crisis and cause another resource crisis, as petroleum has. By international definition, a country with water shortage problems is one with an annual refreshable water amount, per capita, of less than 1000 stere. At present over 100 countries have water shortages and, among these, 40 countries have severe shortages and two billion people lack suitable, drinkable water. Not only arid regions are affected—some European countries and eastern America have plentiful rainfall but suffer from water shortages. Drought and water shortage has led to misery in many areas of the world.

17. Loss of land resources

Land resources could be converted in a short time if there were changes to cultivation systems, such as the use of fertilisers, pesticides, and herbicides; the transformation of economic construction and economic systems; and changes to urbanisation and marketing orientations. Population growth and the improvement of quality of life requires more housing which inevitably occupies cultivated land, forests and grassland. Land use for business, mining, factories and roads is much more economic than for cultivation or grazing, but the irrational use of land resources causes damage to it. Irrational cultivation, overgrazing, deforestation and improper irrigation causes land degradation, soil erosion and desertification. As an example, the Sahara desert has inundated 65 000 square kilometres of arable land during the past 50 years in the Sahel region. In addition, environmental pollution and sea level rises also reduce land resources.

18. Decrease in bioresources

Besides forest decrease and the damage to wildlife habitat, natural grassland (about 30 million square kilometres worldwide) is damaged quantitatively and qualitatively because of climate change and human activity such as overgrazing. Therefore grassland animals are decreasing. Overfishing in oceans also reduces the propagating ability of fish. Wild fauna and flora resources are just one node in the ecosystem, but their destruction endangers human life.

19. Population pressure

Overpopulation has already become a problem for humans. Population growth puts pressure on resources and the environment because of the improvement of living standards, the demand for more resources and social progress. Also, since the average human life span has been prolonged, the increasing numbers of people in the ageing population have produced many social issues.

In 1830, the global population was about one billion. In 1930, the population reached two billion, which means that in just 100 years the population increase was equal to human evolution for several million years. In 1960, the population had reached three billion (over 30 years), in 1975 four billion (15 years) and in 1987 five billion (12 years) (Liu & Li 2001). According to the UN standard, the population of a country or region whose age above 60 accounts for more than 10 per cent of the total population is an aged society. It is predicted that the world will be an aged society by 2010. See Table 4.1.

Table 4.1   The aged trend in the world (per cent)

Year	1950	1960	1970	1980	1990	2000	2010
World	8.1	8.1	8.4	8.6	9.2	9.8	10.7
Developed regions	11.6	12.6	14.5	15.5	17.6	19.2	21.2
Developing regions	6.4	6.1	6.1	6.3	6.9	7.6	8.5

Source: Ma (1999)

Population inflation threatens resources, the environment and even human society itself. Since humans are both producers and consumers, population growth requires more natural resources such as food, housing and water, and more education, hygiene and medical treatment, while at the same time increasing waste to the environment.

20. Sustainable development strategies in the future: Establishing proper environmental ethics

To solve global resource issues, and the environment and population problems mentioned above, changes to ideology and adjustments in ethical relationships between humans and natural resources are needed. It is the minimum moral standard that biodiversity and the natural ecosystem's sustainable subsistence should not be damaged by human activities. People should exert their environmental rights equally without compromising the rights of other people and species, or the ability of future generations to survive and develop. It is necessary to give much more care to natural ecosystems and habitats in order to bring this to fruition. Humans should be regarded as common members of the natural world, and follow the 'rules' set out below.

1   **Adjust the relationship between the human and natural worlds, and advocate harmony and symbiosis.** Humans need to extend environmental rights and obligations to all, and to promote a harmonious relationship between human and human, and between humans and the natural world. Just as in nature, we need to maintain the integrity and health of ecosystems during socioeconomic development to protect those natural factors that affect human survival and development, to utilise natural resources rationally, to prevent environment pollution and ecological breakdown, and to create a high-quality environment for human society to develop. There needs to be a symbiosis and mutual evolution between human and nature.

2   **Respect life, and advocate equality and humanity for all.** All living things need our moral care to protect the already endangered biodiversity on the earth as much as possible, to maintain the ecosystems on which humans and life depend, and make use of species and ecosystems on a continuing basis.

3   **Advocate moderate consumption.** People should be encouraged to consume in an environmentally friendly way so environments and ecosystems can be sustained over a long period. Avoid the luxurious lifestyle, try to consume fewer natural resources (which will improve living standards), and actively implement recycling and clean production methods.

4   **Implement interregional ecological equality and prevent ecological dominance.** We need to consider interregional characters in socioeconomic development; that is, developed regions or regions with superior natural conditions should favour and support developing regions or regions with poor natural conditions. This will prevent the polarisation of regional development, and the interregional transfer of environmental pollution.

21. Establishing an economised social–economic system

Based on the above rules of environmental ethics, developed and developing countries must establish resources based on a national economic system. This system would be characterised by high technological and economic benefits, and less resource consumption and environmental pollution. Such systems include the following features:

- **Rational economic structures and the transformation of economic growth modes**. We should use suitable advanced technologies to improve traditional industries, accelerate hi-tech industry development, devote major efforts to developing service industries, and change the dominant mode of economic growth (high investment, high consumption, high pollution and low efficiency).

- **Highly efficient support systems for energy resources**. Enhance the investment in energy resources technology research, develop important technologies to influence the future direction of energy resources and to improve their utilisation and efficiency, and strengthen energy-saving capabilities.

- **Optimised consumption structure**. Popularise economised technologies in the consumption field, encourage reasonable consumption and encourage the consumption of energy-efficient resource products.

- **New recycling economics**. Promote resources recycling, encourage recycling in enterprises, carry out clean production processes and realise the usefulness of minimising solid waste.

- **Saving energy resources**. Implement standard energy resource efficiency, eliminate high-energy consumption and poor production techniques, establish marketing mechanisms for an energy resources saving system, and form friendly market relationships to save energy resources.

- **Complete programming and cogent policy instruction**. Clarify the dominant objectives of total demand and efficiency of energy resources; set up financial, investment, price and foreign trade policies favourable to energy resource saving; and facilitate the saving and effective use of energy resources.

- **Enact laws and rules, and standardise systems**. Implement related laws, strengthen law enforcement and supervision, establish and implement mandatory standards, and promote energy-efficient resources in production, construction and transportation areas.

- **Distribute information and education**. Carry out various activities to economise energy resources, enhance the consciousness of people (especially young people) of using energy resources wisely, and try to make energy resources saving a conscious action for all citizens.

22. Constructing harmony between humans and the natural world

A society in harmony with human needs and the natural world would be a sustainably developed society. To achieve such a society, some positive measures should be taken:

- **Maintain environmental ethics**. Follow the above rules of environmental ethics, abandon anthropocentric and nature-conquered ideas, and give much more care to nature.

- **Develop the economy and protect the environment simultaneously.** Discard old economic development models which rely on damaging the environment, and develop an economy based on the premise of environmental protection and without compromising future generations.
- **Give attention to both economic and ecological benefits.** Prevent economic benefit from sacrificing ecological benefit, and optimise benefits for economy, ecology and society.
- **Achieve a win-win situation for both productivity development and harmony with nature.** Achieve a harmonious situation between humans and the natural world with growing productivity, wealthy living and a healthy ecosystem.

23. Chapter summary

In this chapter I have developed the idea of what we mean by sustainable development within the global context. Each of the resources in the natural world faces a set of challenges from humans and their continuing need for development and increased consumption of non-renewable energy. To change the direction of past actions, new thinking is needed. There is a clear need for old economies to rethink their practices and develop responses to global needs that match the available resources—for now, and for future generations.

Bibliography

Brown, L & Kane, H 1994, *Full house: Reassessing the earth's population carrying capacity*, Norton, New York [trans. Chen Baiming et al., 1998, Scientific and Technical Documents Publishing House, Beijing].

Environment and Development Center of Chinese Academy of Social Sciences 2001, *Review on Chinese environment and development*, vol. 1, Social Science Documents Publishing House, Beijing.

Fan, B (ed.) 1998, *Chinese natural disasters and disaster management*, Heilongjiang Education Press, Haerbin.

Flavin, C & Lenssen, N 1994, *Power surge: Guide to the coming energy revolution*, Norton, New York [trans. Zhang Kangsheng et al., 1998, Scientific and Technical Documents Publishing House, Beijing].

Huang, W & Niu, Y 1998, 'The sand and dust storm damage and its countermeasure in Northwest China', *Journal of Arid Land Resources and Environment*, vol. 12, no. 3, pp. 83–8.

Investigation Team on National Severe Natural Disasters, 1990, *Natural disasters and reduction*, Seismological Press, Beijing.

IPCC 2001, *Climate change 2001 – The scientific basis. Contribution of Working Group I to the Third Assessment Report of the Intergovernmental Panel on Climate Change*, Cambridge University Press, Cambridge.

IPCC 2007, *Climate change 2007 – The physical science basis. Working Group I contribution to the Fourth Assessment Report of the Intergovernmental Panel on Climate Change*, Cambridge University Press, Cambridge.

Li, R 1998, *Humans and food*, Hubei Science and Technology Press, Wuhan.

Liu, Y et al., 1991, *Chinese population and regional characters*, China Ocean Press, Beijing.

Liu, Y & Li, X (eds) 2001, *Introduction to environmental ecology*, Hunan University Press, Changsha.

Liu, Y & Zhou, H (eds) 2001, *Situation in resources and environment in China and sustainable development*, Economic Science Press, Beijing.

Lu, Y (ed) 1998, *Innovation and future*, Science Press, Beijing.

Ma, Y 1999, *Blue book on human living environment*, Blue Sky Publishing House, Beijing.

Nanjing University et al. (eds) 1983, *Dictionary of geography*, Shanghai Cishu Press, Shanghai.

National Conditions Analyst Team of Chinese Academy of Sciences, 1989, *Survival and development*, Science Press, Beijing.

—— 1997, *Agriculture and development*, Liaoning People's Publishing House, Shenyang.

National Office for Desertification Control 1997, *Chinese desertification report*, China Forestry Publishing House, Beijing.

Postel, S 1997, *The last oasis: Facing water scarcity*, Norton, New York [trans. WU Shaohong et al., 1998, Scientific and Technical Documents Publishing House, Beijing].

UNFCCC 1992, *United Nations Framework Convention on Climate Change*, UN, Geneva, viewed 24 May 2007, (http://unfccc.int/resource/docs/convkp/conveng.pdf).

Wang, T 2005, 'Drawing lessons from other countries to control desert', *Science Times*, 18 May 2005.

—— et al. 2001, 'The situation of dust storms and its strategy in North China', *Bulletin of the Chinese Academy of Sciences*, vol. 5, pp. 343–8.

World Commission on Environment and Development 1987, *Our common future*, Oxford University Press, Oxford.

Yang, Q, Zheng, D & Wu, S 2002, 'Eco-geographic region study in China', *Progress in Natural Sciences*, vol. 12, no. 3, pp. 287–91.

Zhang, X & Zhang, Y 2001, 'Causes, prevention and control of sand storm in North China in recent years', *Journal of Catastrophology*, vol. 16, no. 3, p. 73.

Zhu, Z & Liu, S 1981, *Desertification and delineation in desert regions in North China*, China Forestry Publishing House, Beijing.

Zuo, D (ed.) 1990, *A modern dictionary of geography*, The Commercial Press, Beijing.

Part B:
Case Studies

Eight chapters representing
the diversity of issues in global
geographical locations

5 | Capacity-building for disaster management in southern Thailand

Charlchai Tanavud

1. Southern Thailand in perspective

Southern Thailand, also designated as peninsular Thailand, lies around latitude 13 N and longitude 102 E. The peninsula covers an area of 7 153 917 hectares, which is about one-seventh of the country's total, and has over 2705 kilometres of shoreline. It consists of 14 administrative provinces: Chumphon, Ranong, Phang Nga, Surat Thani, Krabi, Nakhon Si Thammarat, Trang, Phuket, Phatthalung, Satun, Songkhla, Pattani, Yala and Narathiwat.

The terrain is characterised by three mountain ranges, namely the Phuket, the Nakhon Si Thammarat and the Sankala Khiri. The two principal mountain features are the Phuket range on the west of the peninsula, and the Nakhon Si Thammarat range along the north–south axis of the peninsula (see Figure 5.1, page 64). The latter range forms the backbone of this region, with the eastern coastline on the Gulf of Thailand and the western coastline on the Andaman Sea. More coastal plains and stretches of long beaches are found on the east coast, whereas the west coast appears to be a submerged coast, with an irregular shoreline, and many bays and estuaries.

Sediment deposited along the eastern shoreline produces many sandbars and offshore bars. The ridges of the Phuket range are closer to the west coast and contain more granitic cores, providing rich mineral deposits, particularly tin. The landscape between the mountain ranges is mainly made up of low hills and undulating terraces, principally of fluviatile origin (Arbhabhirama et al. 1988). The coastal plains consist of beaches, swamps or brackish alluvium unsuitable for crop production. The river basins of significance are the Tapi, Songkhla Lake, Pak Panang and Pattani basins.

The major soil groups in the peninsula are Paleudults on the undulating and rolling areas, Paleaquults and Tropaquepts on the flat land and coastal plains, and Tropohumods on the beaches. There is evidence of degradation and erosion of soils in southern Thailand. Factors of soil degradation that threaten future soil productivity

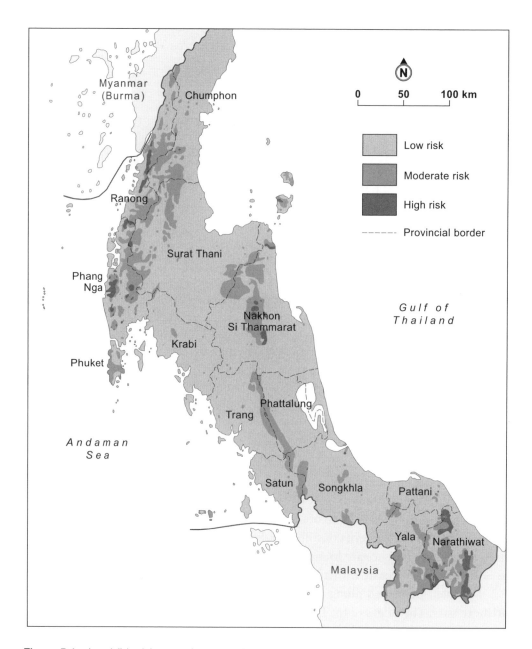

Figure 5.1   Landslide risk map of southern Thailand

include salinisation, acidification, soil nutrient depletion, soil compaction and desertification. Soil erosion is one of the most pressing natural resource problems in the southern region and inevitably results in deterioration of the soil's productive capacity and environmental quality. Land suitable for agriculture in the peninsula totals about 3 506 354 hectares, with 1 129 132 hectares suitable for paddy rice, 2 351 519 hectares for perennial crops and 25 703 hectares for field crops. The remaining 2 673 211 hectares, or 37.8 per cent, is classified as unsuitable for crop production.

The population of the southern region was 8 432 696 in 2004, or 13.6 per cent of the country's total population. On that basis, the population density is 119 per square kilometre. The peninsula, in general, enjoys a tropical climate that provides good moisture and humidity throughout the year. The mean annual rainfall ranges from 4221 millimetres in Ranong Province on the western side of the peninsula to 1658 millimetres in Surat Thani Province across the central mountains on the eastern side. The extreme maximum temperature is 40°C, while the minimum temperature is 9°C. Landslide activity and associated flooding triggered by heavy rainfall are therefore common on steep upland slopes of granitic mountains where forest cover is removed (DeGraff 1989).

The main crop in southern Thailand is rubber, which is grown mostly on sloping lands with good drainage. The cultivation of upland crops is mainly restricted to the deep and well-drained soils found on terraces on the east coast of the peninsula. Broadcast rice is grown in large areas on the east coast of the southern region, on recent marine brackish water and alluvial deposits (Arbhabhirama et al. 1988). Recently, an increasing area along the eastern coast has been devoted to shrimp farming (Tanavud et al. 2001). The peninsular area affords excellent access to the seas, with 12 of the 14 provinces touching the coast. The region's tourism potential is therefore considerable. The long coastlines also make the southern region susceptible to both storm surges induced by typhoons, and tsunamis generated by earthquakes beneath the sea floor.

2. Land use changes

Over the last four decades, the demand for agricultural land in southern Thailand has intensified because of population pressure and a structural shift in the agricultural sector from subsistence farming to commercial farming. In response to the rising demand, farmers have increased their production by expanding cultivable land through encroachment of forests in the upland watershed, rather than increasing yield per unit area. Through the use of satellite imagery and a geographic information system, it was found that, in 1980, forests in southern Thailand covered an estimated 2 632 726 hectares (Table 5.1, page 66), representing the largest land use category in the region (Tanavud, Yongchalermchai & Bennui 2004a). Between 1980 and 1990, forest resources were depleted by a total of 253 731 hectares, equivalent to an annual loss of 25 373 hectares. As a result, by 1990, forest accounted for only 34 per cent of the total land area and had dropped from the largest to the second largest type of land use. During the period from 1990 to 2000, the area under forest cover was further reduced by a total of 334 972 hectares, representing an annual decrease through this decade of 33 497 hectares. Despite the proclamation of a Royal Decree in 1989 for a nationwide ban on commercial logging, the annual average rate of deforestation in southern Thailand was higher during 1990–2000.

Over the 20-year period from 1980 to 2000, the natural forest area in southern Thailand declined by a total of 588 703 hectares, equivalent to a decrease of 22.4 per cent, or an annual decrease of 29 435 hectares (see Table 5.1). In contrast, during the same time frame, the proportion of rubber plantation areas dramatically expanded from 1 354 653 hectares to 2 946 164 hectares, reflecting an annual increase of 79 575 hectares.

Table 5.1   Land use changes in southern Thailand between 1980, 1990 and 2000

Land use category	Area (ha)		
	1980	1990	2000
Forest	2 632 726	2 378 995	2 044 023
Rubber	1 354 653	2 684 188	2 946 164
Rice	1 143 656	779 297	649 335
Fruit trees and mixed orchards	411 628	698 056	829 891
Aquaculture	—	—	84 528
Urban and built up land	15 692	47 640	139 543
Miscellaneous land	1 438 889	409 068	303 760
Total	6 997 244	6 997 244	6 997 244

Source: Tanavud, Yongchalermchai & Bennui (2004a)

The dramatic increase in the price of rubber during this time period and the subsidies provided by the Thai Government's Office of Rubber Replanting Aid Fund have been instrumental in promoting land conversion to rubber plantation (Usher 1994). These direct causes of deforestation, however, obscure the underlying causes, which include poverty, inequitable resource tenure and population pressure. It is noteworthy that Thailand is one of the three countries that have the highest deforestation rate in the South-East Asian region (UN 2000). The extensive deforestation and its conversion to rubber plantations have inevitably resulted in a change in hydrological conditions, thereby increasing the risk of erosion hazards.

There were 1 904 177 hectares of the peninsula's land surface area designated as headwater source areas, where permanent forest cover must be maintained for such purposes as watershed protection (Table 5.2). The criteria for the classification of the headwater source areas are based primarily on physical characteristics of landscape units, namely slopes, elevation, landform, soil and geology. For the most part, these forested watershed areas are located at higher elevations, with steep slopes and erosive landforms. Tanavud, Yongchalermchai and Densrisereekul (2007) assessed the status of forests in the headwater source area in Songkhla Lake Basin, southern Thailand, and found that only 64 336 hectares, or 59.7 per cent, of the basin's headwater source areas remains under forest cover (Table 5.2). The encroachment and conversion of forest to rubber plantation in these environmentally critical areas has contributed to a reduction in the resilience of the natural upland ecosystems. As a consequence, southern Thailand has become highly vulnerable to natural disasters, primarily floods and landslides (UNEP 1994).

Table 5.2   Headwater source area and forested area in Songkhla Lake Basin

Songkhla Lake Basin	Areas (ha)	Forested areas in headwater source areas	
		ha	%
Watershed class 1	63 522	46 528	73.2
Watershed class 2	44 233	17 808	40.3
Headwater source areas (classes 1 and 2)	107 755	64 336	59.7

Source: Tanavud, Yongchalermchai & Densrisereekul (2007)

It was not until the early 1980s that shrimp cultures for black tiger shrimp (*Penaeus monodon*) were introduced by the government as a means of providing nutrition, improving household incomes and enhancing employment opportunities for southern Thailand's population. Thus, the coastal areas devoted to aquaculture dramatically increased from 1980 to 2000 (Table 5.1). This increase came from the conversion of mangrove areas along the coastal areas, resulting in a degradation of coastal ecosystems. Consequently, the coastal zones of southern Thailand have become highly vulnerable to extreme natural events. Due to outbreaks of disease along the coast and the development of low salinity culture techniques for shrimp cultivation, new cultivation areas have emerged along the estuaries of the main rivers and/or canals some distance upstream from the coast. The untreated waste and salty water disposed from these newly established shrimp farms pollutes neighbouring land areas, resulting in an abandonment of cultivable land, particularly rice fields (Tanavud et al. 2001).

In response to the phenomenal growth of the tourism and aquaculture industries, beach resorts, tourism facilities and aquaculture infrastructure have replaced much of the mangrove and beach forests along the coastal shores (Tanavud, Yongchalermchai & Densrisereekul 2006). Moreover, many of the coastal sand dunes that act as natural barriers against extreme events have been removed to make way for the construction of beach resorts, walkways and roads. As a consequence of the destruction of the coastal ecosystem structure, coastal areas of southern Thailand have become extremely vulnerable to natural calamities, particularly typhoons and tsunamis. Ecologically poor land use practices, both on the uplands and along the coastal areas of the peninsula, have set the stage for the occurrence of natural disasters triggered by torrential rains or earthquakes beneath the ocean floor.

3. Consequences of land use changes

3.1 Erosion

The dramatic depletion of southern Thailand's forest resources has inevitably led to the occurrence of soil erosion, with consequent impacts on soil productivity and environmental quality. Based on the Universal Soil Loss Equation (USLE), it was estimated through the use of remotely sensed data and geographic information systems that 125 720 000 tonnes of soil were lost from southern Thailand annually (Table 5.3). This represented an average total soil loss of 17.9 t ha^{-1} y^{-1}. Some 6 119 677 hectares, or 87.5 per cent of the total land area, exhibited a very slight rate of soil erosion. Approximately 164 533 and 36 618 hectares, equivalent to 2.3 and 0.5 per cent of the total land area, were subjected to severe and very severe erosion respectively (Tanavud et al. 2002). Most of these severely eroded lands were found on the mountainous terrain to the north-west of the southern region.

It was observed that, of the 14 provinces, Ranong and Chumphon Provinces were the highest and second highest areas subjected to very severe rates of soil erosion (Table 5.4). Severity of erosion in these two provinces could be attributed to high rainfall, mountainous terrain and inappropriate land use. It is noteworthy that Phatthalung and Pattani Provinces had no areas subject to very severe rates of soil erosion (Tanavud et al. 2002).

Table 5.3   Soil losses from different erosion classes in southern Thailand

Erosion Classes	Rating (t ha^{-1} y^{-1})	Soil loss (t y^{-1})	Area		Average (t ha^{-1} y^{-1})
			ha	%	
Very slight	< 6.25	3 034 000	6 119 677	87.50	0.50
Slight	6.25–31.25	9 676 000	422 796	6.04	22.89
Moderate	31.25–125	16 180 000	250 629	3.58	64.56
Severe	125–625	53 110 000	164 533	2.35	322.79
Very severe	> 625	43 720 000	36 618	0.52	1193.95
Total		125 720 000	6 994 253	100.00	

Source: Tanavud et al. (2002)

Table 5.4   The distribution of erosion-affected areas in the provinces of southern Thailand

Provinces	Erosion hazard areas (ha)									
	Very slight	%	Slight	%	Moderate	%	Severe	%	Very severe	%
Nakhon Si Thammarat	867 277	87.2	48 026	4.8	30 292	3.0	44 698	4.5	4 217	0.4
Krabi	439 617	92.6	19 582	4.1	12 611	2.7	2 717	0.6	286	0.1
Phang Nga	336 742	86.9	20 301	5.2	23 241	6.0	4 513	1.2	2 601	0.7
Phuket	42 116	78.8	2 531	4.7	6 841	12.8	1 911	3.6	34	0.1
Surat Thani	1 164 966	89.9	92 517	7.1	16 302	1.3	21 141	1.6	437	0.03
Ranong	235 426	74.5	46 207	14.6	3 454	1.1	11 647	3.7	19 079	6.0
Chumphon	527 298	89.9	23 514	4.0	5 991	1.0	21 415	3.6	8 424	1.4
Songkhla	639 496	87.8	45 771	6.3	17 235	2.4	26 232	3.6	30	0.04
Satun	246 178	96.0	3 505	1.4	6 025	2.3	783	0.3	15	0.05
Trang	429 541	90.0	24 265	5.1	8 959	1.9	8 546	1.8	635	0.1
Phatthalung	321 791	96.2	6 467	1.9	6 315	1.9	—	—	—	—
Pattani	186 796	93.6	6 839	3.4	5 265	2.6	680	0.3	—	—
Yala	325 493	72.4	64 192	14.3	44 343	9.9	15 100	3.4	189	0.04
Narathiwat	356 939	80.1	19 081	4.3	63 755	14.3	5 152	1.2	672	0.2
Total	6 119 677	87.5	422 796	6.0	250 629	3.6	164 533	2.4	36 618	0.5

Source: Tanavud et al. (2002)

3.2 Disasters

Since people's dependence on land is so great and enforcement procedures for sustainable use of land are so weak, land use practices which are ecologically poor continue, both in the headwater source areas and on the coastal areas. As a result, southern Thailand has been under constant threat and at increasing risk of future catastrophic events. Consequently, the 1988 landslides were inevitable when southern Thailand experienced a major storm event (Tanavud et al. 2000), the devastating 2000 floods were unavoidable when Hat Yai City recorded 270 millimetres of torrential rainfall over a five-day period (Tanavud et al. 2004), and the 2004 tsunami catastrophes were inescapable when a massive earthquake of 9.3 magnitude occurred beneath the sea floor off the western coast of northern Sumatra, Indonesia (Tanavud, Yongchalermchai & Densrisereekul 2006).

3.2.1 Landslides

The catastrophic landslide of 20–23 November 1988 in southern Thailand has been described as one of the worst natural disasters in Thailand's recorded history. Estimates of damage to property, crops and infrastructure range from US$160 to US$180 million (Rau 1991). The disaster centred in Phipun Basin, where 175 people perished. Logs and landslides swept down from the mountainsides surrounding the basin. They destroyed homes and buried the valley's rice fields and fruit orchards with soil materials developed from the weathering of granitic bedrocks on steep mountains. Dramatic front-page newspaper stories and television reports led the government to impose a nationwide ban on commercial logging—an action unprecedented anywhere in the world (Tanavud 1993). The landslide catastrophe was attributed to an extensive conversion of natural forest to rubber plantation on the upland watershed areas, and triggered by intense rainfall (Rau 1991).

3.2.2 Floods

The unprecedented flooding of Ha Yai, southern Thailand, from 21 to 24 November 2000, which was triggered by torrential rains, has also been described as one of the worst natural disasters in the history of urban Thailand. In this tragic event, 30 people perished, and public utilities, critical facilities and commercial and industrial establishments were severely damaged or destroyed. Estimated total economic losses exceeded US$220 million. Indirect losses in terms of lost production and the cost of economic recovery make the estimation even higher (Tanavud et al. 2004).

3.2.3 Tsunamis

On 26 December 2004, an earthquake measuring 9.3 on the Richter scale ripped apart the sea floor off the western coast of northern Sumatra, Indonesia. The sudden vertical rise of the seabed by several metres during the quake displaced massive volumes of water, resulting in a devastating tsunami. This seismic sea wave travelled thousands of kilometres across the Indian Ocean, and ravaged the Andaman coast of southern Thailand at 9.30 a.m. local time. Six coastal provinces— Phuket, Phang Nga, Krabi, Ranong, Trang and Satun—have been severely affected by the tsunami tragedy, and suffered a total of over 5395 deaths, more than half of whom were foreign tourists, with another 2822 reported missing (Department of Disaster Prevention and Mitigation 2005). Of the six affected provinces, Phang Nga was the worst hit, with some 4224 lives lost and 7003 hectares of land area devastated. Takua Pa District, which was a prime tourist area with numerous beach resorts, was the most brutally affected area in Phang Nga Province. The maximum run-up height of 10–12 metres was recorded at this district, and the trail of destruction left by the devastating wave at this locality extended up to a kilometre inland (Tanavud, Yongchalermchai & Densrisereekul 2006). As a result, Takua Pa District suffered a considerable loss of human lives and livelihoods, as well as damage to settlements, infrastructure, tourism facilities, fragile ecosystems and protected areas, including coral reefs, and habitats of endangered and vulnerable species, on a scale never before seen in Thailand.

These calamitous events present a significant threat to southern Thailand's development strategies by destroying lifeline infrastructure, critical facilities and production capacity, interrupting economic activity, and creating irreversible changes in the natural resource base. It was estimated that losses due to these extreme events each year offset the aid provided for Thailand by developed countries. Scarce resources that were earmarked for development projects in Thailand are therefore diverted to relief and reconstruction following disasters, thus impeding the process of sustainable development.

4. Hazard and risk assessments

To identify the nature and extent of the landslide and flood problems, and to design and implement appropriate measures to reduce the risk of future disasters in southern Thailand, areas that are prone to landslide and flood occurrences have been mapped. These maps provide information on disaster vulnerability of an area of concern before any development projects are undertaken. They also facilitate identification of safe places for the evacuation of people in an emergency. Based on the causative factors that contribute to flooding—rainfall, the basin's side slopes, gradient of main drainage, channel drainage density, land use, soil types and road and/or railways—it was assessed through the use of satellite imagery and GIS techniques that 209 100 hectares, 762 900 hectares and 343 100 hectares of land in southern Thailand are subjected to low, moderate and high flood risks respectively (Table 5.5, page 72). It is worth noting that 143 700 hectares subject to high flood risk, equivalent to 14.4 per cent of the total land area of southern Thailand, was found in Nakhon Si Thammarat Province.

These hazard zones and areas at risk present a significant threat to the country's development strategies in the years to come. With continuing forest encroachment in the headwater source areas, and replacement of mangroves and beach forests by tourism facilities and aquaculture structures along the coastal shores, landslides, floods and tsunamis are predicted to be recurrent phenomena in southern Thailand in the future. Losses due to these natural events in the future will deprive southern Thailand of financial resources that should be used for economic and social development. Enhancing the capacity of the communities at risk to save lives and limit the impacts of the disasters is the most critical element for the long-term stability and sustainable development of southern Thailand.

Based on geological and biophysical factors that give rise to landslides—such as bedrock, slopes, land use and rainfall—it was assessed through the use of satellite

Table 5.5 Distribution of areas facing flood risk in provinces of southern Thailand

Province	Area (ha)	Risk levels					
		High		Moderate		Low	
		Area (ha)	%	Area (ha)	%	Area (ha)	%
Nakhon Si Thammarat	996 700	143 700	14.4	165 600	16.6	10 100	1.0
Krabi	485 400	4 000	0.8	65 100	13.4	200	—
Phang Nga	394 400	8 100	2.1	74 600	18.9	—	—
Phuket	56 300	1 600	2.8	—	—	—	—
Surat Thani	1 312 600	22 500	1.7	90 800	6.9	57 500	4.4
Ranong	319 200	5 300	1.7	32 600	10.2	800	0.3
Chumphon	589 500	12 600	2.1	23 000	3.9	65 300	11.1
Songkhla	789 300	21 400	2.7	36 100	4.6	25 800	3.3
Satun	261 700	10 600	4.1	72 300	27.6	15 000	5.7
Trang	476 100	24 400	5.1	56 400	11.8	3 900	0.8
Phatthalung	375 100	18 900	5.0	25 200	6.7	8 500	2.3
Pattani	200 400	26 500	13.2	36 800	18.4	19 400	9.7
Yala	450 000	6 700	1.5	17 800	4.0	2 600	0.6
Narathiwat	447 100	36 800	8.2	66 600	14.9	—	—
Total	7 153 800	343 100	4.8	762 900	10.7	209 100	2.9

Source: Tanavud, Yongchalermchai & Bennui (2004b)

Table 5.6   Distribution of areas facing high landslide hazard and high landslide risk in each province of southern Thailand

Province	High hazard		High risk	
	Area (ha)	%	Area (ha)	%
Narathiwat	127 734.95	33.73	186 053.30	33.58
Nakhon Si Thammarat	70 897.94	18.72	109 310.74	19.73
Phang Nga	86 047.87	22.73	110 044.07	19.86
Ranong	45 592.67	12.04	76 865.36	13.87
Yala	38 434.62	10.15	54 318.97	9.81
Surat Thani	3 295.51	0.87	5 432.61	0.98
Pattani	1 148.73	0.30	5 068.00	0.91
Chumphon	2 389.07	0.63	3 142.43	0.57
Trang	1 643.71	0.43	1 588.65	0.29
Songkhla	469.52	0.13	1 321.80	0.24
Krabi	297.37	0.08	361.09	0.07
Phatthalung	132.40	0.04	328.13	0.06
Satun	234.33	0.06	7.65	0.01
Phuket	344.04	0.09	128.20	0.02
Total	**378 662.73**	**100.00**	**553 971.00**	**100.00**

Source: Tanavud et al. (2000)

imagery and GIS techniques that there are 553 971 hectares, equivalent to 7.9 per cent of the total land area of southern Thailand, facing high risk from landslides (Table 5.6). Of the 14 provinces in southern Thailand, Narathiwat and Nakhon Si Thammarat contain a total of 295 364 hectares, equivalent to 53.3 per cent of the high risk areas (Table 5.6).

5. Disaster management

While the occurrence of the catastrophic events cannot be prevented, the magnitude of their impacts in terms of loss of life and destruction of property could be minimised through effective disaster management. Disaster management is widely seen as embracing four elements: mitigation, preparedness, response and recovery. In Thailand, however, disaster management focused primarily on the emergency period response and post-impact recovery. In recent years, considerable structural mitigation activities have been undertaken to reduce the effects of the disasters on lives, property and production in southern Thailand. However, the occurrence of a recent series of natural calamities such as floods, landslides and tsunamis have indicated that such approaches are not sufficient. In recognition of this fact, much greater emphasis must be placed on preparedness approaches to ensure effective and adequate response to the impacts of disasters. Examples of preparedness measures include issuing timely and effective early warnings and removing people and property from a threatened location.

Most of the casualties and fatalities due to past natural disasters in southern Thailand can be attributed to the government's failure to warn vulnerable communities of the imminent arrival of a catastrophe. In the aftermath of the December 2004 tsunami disaster, the Thai National Disaster Warning Center (NDWC) was officially established in May 2005, five months after the tsunami event. The NDWC, which has the lead responsibility for issuing disaster alerts in Thailand, receives advisory information on the severity, location and the likelihood of disasters from the Thai Meteorological Department as well as overseas seismological stations including the Pacific Tsunami Warning Centre, the Japan Meteorological Agency, the United States Geological Survey, and the National Oceanic and Atmospheric Administration (NOAA). The centre will analyse the information obtained according to their standard operating procedures, and deliver warning messages to the public through various communication media such as radio, television, mobile phones and siren towers.

However, it is still questionable as to how effective this would have been in alerting the public of the impending danger in southern Thailand where communication systems are inadequate, and awareness of the dangers posed by natural disasters is low. The lack of disaster education and precautionary behaviour of the vulnerable populations, even when observing a prolonged period of heavy rains that gave an indication of the impending flash floods and/or landslides, or the suddenly receding sea water that provided a signal of the approaching tsunami waves, also contributed to the massive loss of life and damage to property. There is therefore an urgent need to strengthen and improve southern Thailand's disaster response capacity to assist, enhance and assure sustainable development.

6. Capacity-building

One of the critical parts of the Thai Government's efforts to enable vulnerable communities to cope better with the impacts of natural calamities is capacity-building, and this has become one of Prince of Songkhla University's primary roles. The specific activities undertaken are detailed below.

In the aftermath of the 2004 tsunami, remarkable achievements have been made in the field of disaster education, especially in higher levels of education. Prince of

Songkhla University has pioneered the incorporation of disaster risk reduction as part of the university curriculum. A disaster management subject has been offered to both undergraduate and postgraduate students at the Faculty of Natural Resource. This subject, which includes sustainable development concepts, focuses on all aspects of disaster risk management, namely mitigation, preparedness, response and recovery. The subject covers topics such as hazard identification and assessment, disaster cycles, vulnerability and risk analyses, disaster and environment, community based approaches to disaster risk management, using remotely sensed data and GIS for disaster management, and disaster and sustainable development. A field visit to disaster-affected areas is also an essential part of the course. To complete the subject, the students are required to prepare an assignment on disaster-related topics, write a report after returning from the field trip, and take a final examination at the end of the course. After the December 2004 tsunami event, this subject was offered to the students during the first and second semesters of 2005 as well as during the first semester of 2006. Approximately 15–20 undergraduate and postgraduate students enrolled for the subject in each semester. This disaster education is recognised as a long-term strategy to provide protection from future natural disasters.

Training has been organised for relevant local government officials and sub-district administration organisation personnel who have responsibility for disaster risk reduction. This training enhances their capacity and skills in planning for, and minimising the impact of, natural disasters. It includes procedures for warning and notification, evacuation and assembly locations, search and rescue operation, emergency response management, relief and rehabilitation, mitigation and preparedness, and damage and needs assessment. In addition, a technical training course is also conducted for all response and recovery personnel in disaster-prone areas on topics such as inundation and numerical modelling, GIS-based hazard and vulnerability mapping, risk assessment, disaster-resistant construction techniques, updating coastal bathymetric and geomorphology databases for modelling, sea-level data sampling, and mitigation and preparedness priorities.

Prince of Songkhla University organised a national conference on disaster management from 26–28 December 2005 in Phuket, a tsunami-affected province. The conference's objective was to offer a forum for interaction and networking between academics, local government agencies, sub-district administrative organisations, non-government organisations, and disaster-affected communities to share and exchange knowledge, experiences, expertise and lessons learned on issues relating to disaster risk reduction. The conference encompassed lectures, scientific sessions with oral and poster presentations, panel discussions and field excursions to the tsunami-affected areas. It was a highly successful event with over 100 participants. The conference proceedings—a compilation of the papers presented at the conference—were published after the conference to ensure that the information reaches a larger segment of society.

Prior to the 2004 tsunami catastrophe, the Moken—the seafarers of the Andaman Sea in southern Thailand—did not need sophisticated equipment to warn them of an impending danger from the sea. For them, there were many natural warning signs leading up to the December 2004 tsunami. Three days before the tsunami struck, deep sea creatures appeared near the surface: crabs and mantis shrimp left their burrows to migrate elsewhere, and fish became highly agitated, leaping out of the water in entire schools. These telltale signs, together with ancestral spirits, remind the indigenous Moken of impending tsunami waves. The final sign—the rapidly receding sea water which arrived on the morning of 26 December 2004—informed the Moken to flee the coastal areas to higher ground before the sea water came surging back in killer waves. With the orally transmitted knowledge of tsunami sense they have passed down through their legends, only one Moken, who had suffered from partial paralysis for several years, was unable to reach high ground in time and thus lost his life to the powerful waves (Tanavud 2005). This indigenous knowledge and local wisdom related to the natural disasters of seafaring communities such as the Moken and the Urak Lawoi has since been promoted.

Public awareness programs have been developed to provide information and knowledge to vulnerable communities on the nature and processes of the hazard that threatens their community and how to respond. Brochures, posters, calendars, and announcements on radio and television are also used to make the local communities aware of past disaster events, their vulnerabilities and their disaster roles and responsibilities, and instil a culture of safety against natural hazards. This awareness is essential for motivating people at risk to become more active in vulnerability reduction activities, and for stimulating local communities to assume more responsibility for their own safety.

Research on topics related to hazard identification, vulnerability analysis and risk assessment have been undertaken. Moreover, research projects are carried out on topics such as the effects of the 2004 tsunami on coastal ecosystems, inventory of historical inundation data and run-up data, the use of remote sensing and GIS technology in producing data layers (including critical infrastructure, bathymetry and digital elevation models), and assessment of tsunami vulnerability in the coastal areas. More collaboration between international research communities to pool resources, scientific knowledge and expertise are encouraged to bring about an increase in the capacities of the relevant authorities and vulnerable communities to cope effectively with such eventualities.

The catastrophic events such as those experienced in the 1988 landslides, the 2000 floods and the 2004 tsunami in southern Thailand have served as tragic reminders of the long-term price which must be paid for unsustainable development. It is anticipated that the successful implementation of the capacity-building options described above will make a significant contribution to the resilience of the communities to adverse natural phenomena—the resilience that is so crucial to sustainable development in southern Thailand.

Bibliography

Arbhabhirama, A, Phantumvanit, D, Elkington, J & Ingkasuwan, P 1988, *Thailand natural resources profile*, Oxford University Press, Oxford.

DeGraff, JV 1989, 'Landslide activity resulting from the November 1988 storm event in southern Thailand and associated resource recovery needs', in *Safeguarding the future: Restoration and sustainable development in the South of Thailand*, National Operation Center, NESDB & USAID, Bangkok, pp. 153–95.

Department of Disaster Prevention and Mitigation 2005, 'Tsunami disaster situation', Ministry of Interior, Bangkok, 27 May 2005.

Rau, JL 1991, 'Assessment and mitigation of landslides in rural areas of Southeast Asia', in *Disaster management and regional development planning with people's participation*, vol. 2, United Nations Centre for Regional Development, Nagoya, Japan, pp. 210–35.

Tanavud, C 1993, 'Reclamation of land affected by devastating floods, I. The nature of the soil materials and defining the problems', *Thai Journal of Soil and Fertilizers*, vol. 15, pp. 108–17.

—— 2005, 'Report on the impact of the December 2004 tsunami on the Andaman coast of southern Thailand' in Tanavud, C, Tamura, T, Miyaki, T & Kashima, K (eds), *Proceedings of the International Conference on Environmental Hazards and Geomorphology in Monsoon Asia: December 2004*, Hat Yai, Thailand, pp. 153–61.

——, Bennui, A & Sansena, T 2006, 'Study of forest status in headwater source area in southern Thailand using remote sensing and GIS', *Geographical Journal*, vol. 30 (in press).

——, Yongchalermchai, C & Bennui, A 2004a, 'Land use changes and its consequences in southern Thailand', *in Proceedings of the 4th International Congress of the European Society for Soil Conservation, 25–29 May 2004*, Budapest, pp. 13–15.

——, Yongchalermchai, C & Bennui, A 2004b, 'An assessment of flood risk in southern Thailand', *Journal of Remote Sensing and GIS Association of Thailand*, vol. 5, no. 3, pp. 10–21.

——, Yongchalermchai, C, Bennui, A & Densrisereekul, O 2001, 'The expansion of inland shrimp farming and its environmental impacts in Songkhla Lake Basin', *Kasetsart Journal (Natural Science)*, vol. 35, pp. 326–43.

——, Yongchalermchai, C, Bennui, A & Densrisereekul, O 2004, 'Assessment of flood risk in Hat Yai Municipality, Southern Thailand, using GIS', *Journal of Natural Disaster Science*, vol 26, no. 1, pp. 1–14.

——, Yongchalermchai, C, Bennui, A & Navanugraha, C 2000, 'Application of GIS and remote sensing for landslide disaster management in Southern Thailand', *Journal of Natural Disaster Science*, vol. 22, no. 2, pp. 67–74.

——, Yongchalermchai, C & Densrisereekul, O 2006, 'Effects of the December 2004 tsunami and disaster management in Southern Thailand', *Science of Tsunami Hazards*, vol. 24, no. 3, pp. 206–17.

——, Yongchalermchai, C & Densrisereekul, O 2007, 'Assessing the impacts of the implementation of a watershed classification system on land use practices, water and soil resources in Songkhla Lake Basin', *Journal of Remote Sensing and GIS Association of Thailand*, vol. 8, no. 1, pp. 31–43.

——, Yongchalermchai, C, Navanugraha, C & Bennui, A 2002, 'Erosion hazard assessment in Southern Thailand', *Journal of Remote Sensing and GIS Association of Thailand*, vol. 3, no. 3, pp. 1–14.

UN 2000, *State of the environment in Asia and the Pacific*, United Nations, New York.

UNEP 1994, *Strengthening disaster management strategies in Thailand*, Asian Disaster Preparedness Center (ADPC) & Asian Institute of Technology (AIT), Bangkok.

Usher, AD 1994, 'After the forest', *Thai Development Newsletter*, vol. 26, pp. 20–32.

6 | Water-saving in a city: A case study of Beijing City

WU Shaohong, Zhang Guoyou, Chen Yuansheng and Yiin Yunhe

1. Background

Beijing City is located in the northern plain area of China, with a temperate monsoon climate of humid and sub-arid periods each year. The city is the political and cultural centre of China. There are more than 14 million permanent residents in the city and a floating population of more than one million. Water shortage is a serious problem, which constrains sustainable socioeconomic development. The average water resource available for each Chinese person is only 2500 cubic metres, which is a quarter of the world's average. However, the water resources for the residents of Beijing are less than 300 cubic metres per capita, which is less than an eighth of the national average, well below the lower limitation of the international standard of 1000 cubic metres. Beijing is therefore a megacity with a serious water shortage problem. The water resources available in Beijing are 3.58×10^9 cubic metres. The available groundwater resources are 2.5×10^9 cubic metres annually, but net pumping each year is 2.5993×10^9 cubic metres from five water systems (Table 6.1, page 80).

Groundwater resources are 3.951×10^9 cubic metres (Table 6.2, page 81). Because of the socioeconomic development in the city, large amounts of groundwater are exploited annually, and account for about 2.6×10^9 cubic metres (Table 6.3, page 81). Overexploitation has led to a decrease in groundwater, a trend that is continuing (Figure 6.1, page 82). Accordingly, groundwater levels have changed dramatically (Figure 6.2, page 82).

The Chinese Government started a project to bring water from the south (Changjiang River Basin) to the north in order to alleviate the water shortage situation in Beijing. However, the Beijing public had not been made aware of water-saving methods, so the Beijing Government set up regulations to limit water consumption by citizens. As an example, monthly water consumption that exceeds a defined amount is charged more per unit than the normal price. The new generation, mostly one-child families, do not seem to be taking this issue seriously, so it is important to raise awareness of water-saving in the daily lives of the younger generation.

Table 6.1   Water resources in different water systems in Beijing

Items	Water systems	Daqinghe River	Yongdinghe River	Beiyunhe River	Chaobaihe River	Jiyunhe River	Total
Mountain areas	Area (km²)	1615	2491	1000	4605	689	10 400
	Annual flux (10^9 m³)	0.3100	0.2926	0.1522	0.9013	0.1612	1.8173
	Account for (%)	11.9	11.3	5.9	34.7	6.2	70
	Annual run-off (10^3 m³)	192	118	152	196	234	175
Plain areas	Area (km²)	604	677	3423	1008	688	6400
	Annual flux (10^9 m³)	0.0651	0.04868	0.4782	0.1216	0.0685	0.7820
	Account for (%)	2.5	1.8	18.4	4.7	2.6	30.1
	Annual run-off (10^3 m³)	108	720	140	121	100	122
Total	Area (km²)	2219	3168	4423	5613	1377	16 800
	Annual flux (10^9 m³)	0.3751	0.3412	0.6304	1.0229	0.2297	2.5993
	Account for (%)	14.4	13.1	24.3	39.4	8.8	100
	Annual run-off (10^3 m³)	169	108	143	182	167	155

Source: Zheng & Lu (2001)

Table 6.2   Groundwater resources of Beijing (billion cubic metres per annum)

Areas \ Items	Mountain areas	Plain areas	Repetition of mountain and plain	Total
Downtown area	0.356	0.577	0.211	0.722
Tongzhou		0.178		0.178
Daxing		0.320		0.320
Changping	0.128	0.320	0.108	0.340
Fangshan	0.267	0.440	0.123	0.584
Shunyi	0.008	0.262	0.008	0.262
Miyun	0.273	0.298	0.055	0.516
Huairou	0.307	0.152	0.032	0.427
Pinggu	0.154	0.258	0.126	0.286
Yanqing	0.221	0.162	0.061	0.322
Annual average	**1.714**	**2.967**	**0.724**	**3.957**

Source: Zheng & Lu (2001)

Table 6.3   Annual groundwater exploitation in Beijing (billion cubic metres per annum)

Areas \ Items	Mountain areas	Plain areas	Total
Downtown area	0.015	0.605	0.62
Tongzhou		0.190	0.190
Daxing		0.280	0.280
Fangshan	0.046	0.290	0.336
Changping	0.016	0.220	0.236
Yanqing	0.010	0.090	0.100
Shunyi		0.430	0.430
Miyun	0.037	0.070	0.107
Huairou	0.024	0.080	0.104
Pinggu	0.030	0.200	0.230
Annual average	**0.178**	**2.455**	**2.633**

Source: Zheng & Lu (2001)

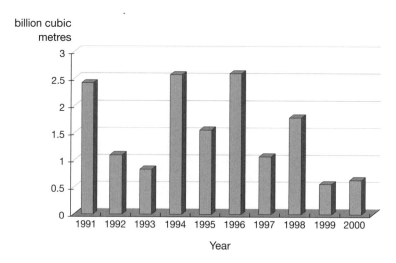

Figure 6.1   Surface water in Beijing, 1991–2000

Source: Zhu & Shao (2003)

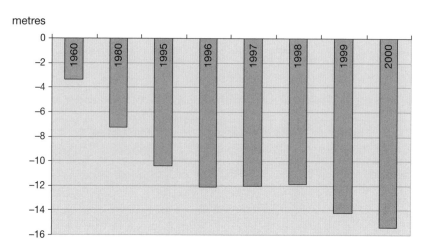

Figure 6.2   Changing groundwater levels in the plains regions of Beijing, 1960–2000

Source: Zhu & Shao (2003)

2. The status of water consumption in Beijing

2.1 The supply and use of water in Beijing

The average annual water supply in Beijing from 1996–2003 was 3.941 billion cubic metres. Of that water supply, 1.295 billion cubic metres were from surface water and 2.619 billion cubic metres from groundwater. Other sources accounted for 0.028 billion cubic metres. The total water use was 3.933 billion cubic metres, of which:

- agriculture took 1.699 billion cubic metres, or 43.2 per cent of the total
- industry took 0.99 billion cubic metres, or 25.4 per cent of the total
- daily consumption took 1.235 billion cubic metres, or 30 per cent of the total.

Table 6.4 and Figure 6.3 show the details.

Table 6.4   Water consumption in recent years (billion cubic metres)

	1996	1997	1998	1999	2000	2001	2002	2003
Surface water supply	1.582	1.501	1.488	1.495	1.325	1.17	0.965	0.834
Groundwater supply	2.731	2.585	2.552	2.676	2.715	2.723	2.424	2.542
Other supply	0.008	0.01	0.007				0.073	0.125
Total supply	**4.321**	**4.096**	**4.047**	**4.171**	**4.04**	**3.893**	**3.462**	**3.501**
Agriculture consumption	1.968	1.812	1.739	1.845	1.649	1.74	1.545	1.292
Industrial consumption	1.264	1.1	1.084	1.056	1.052	0.918	0.754	0.765
Daily consumption	1.089	1.114	1.224	1.27	1.339	1.235	1.163	1.444
Total consumption	**4.321**	**4.026**	**4.047**	**4.171**	**4.04**	**3.893**	**3.462**	**3.501**

Source: The Bureau of Beijing Hydrology (2004)

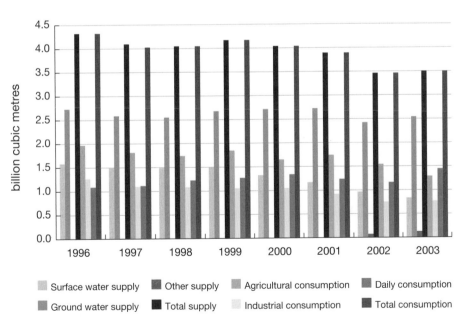

Figure 6.3   Water supply and use in Beijing, 1996–2003

Source: The Bureau of Beijing Hydrology (2004)

In recent years the available water resources have gradually declined. For example, in 2004 the available water in that year was 2.90 billion cubic metres but consumption during the year was 3.46 billion cubic metres. The total water shortage was therefore 0.56 billion cubic metres. By using 0.20 billion cubic metres of recycled water, overexploitation of groundwater was as much as 0.36 billion cubic metres. Table 6.5 and Figure 6.4 show the details of water used in 2004.

Table 6.5 Water used in Beijing in 2004 (billion cubic metres)

Items		Consumption
Agricultural		1.35
Industrial		0.77
Daily	Public service	0.71
	Civil	0.57
Environmental purpose		0.06
Total consumption		**3.46**

Source: Beijing Water Saving Management Office (2005)

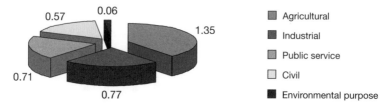

- Agricultural
- Industrial
- Public service
- Civil
- Environmental purpose

Figure 6.4 The composition of water use in Beijing, 2004

Source: Beijing Water Saving Management Office (2005)

2.2 How water resources are being wasted

In daily life

There are over 14 million people in Beijing. Water consumption for the daily lives of its people takes 37 per cent of the total resource, and this is increasing by 4.4 per cent annually. However, leaking water is a serious issue in the city and in 2003 accounted for as much as 17 per cent of the total resource. One observation showed that a leaking tap would waste from 1 to 6 cubic metres per month and a leaking toilet from 3 to 25 cubic metres a month. If there are as many as 600 000 leaking taps and 200 000 leaking toilets, this will waste hundreds of millions of cubic metres of water annually. There seems to be a lack of recognition about this problem of waste despite the fact that water consumption is increasing very quickly: daily water consumption per capita was 75 litres in 1959 but is nearly 200 litres nowadays. Some of this is wasted, through practices such as taking several showers a day (with 200–300 litres of water being used each time), using multi-flushing toilets (with older toilets using 13 litres of water each time), and using a washing machine without a full load. Compared to other cities of China and even other countries, the daily water consumption of 285 litres per person in Beijing is definitely too high (Table 6.6).

Table 6.6   Water consumption in other cities and countries

Country	City	Water consumption (litre/capita/day)	Year
China	Guangzhou	280	1999
	Nanjing	120	1999
	Shenzhen	160	1999
	Changchun	120	1999
	Tianjin	100	1999
	Taipei	188	1997
	Hong Kong	213	1996
Japan	Tokyo	190	1998
Germany	Berlin	117	1999
	Frankfurt	171	1999
Denmark	Copenhagen	159	1990
Australia	Sydney	212	1985
USA	Los Angeles	308	1996
	Philadelphia	341	1996

Source: Beijing Water Saving Management Office (2002)

Since the 1970s, most of the families in China are one-child families. This new generation enjoys rich resources from nature and society, but saving resources is not a conscious activity. This is an area of concern which should be addressed.

Green belt irrigation

According to the Beijing Garden Management Bureau, remote-sensing surveys show that the average green belt area for a Beijing resident is 36.08 square metres at present. It is planned that in five years, forest coverage will reach 70 per cent on hills and mountains and 25 per cent on the plains. The average area of green belt for a

Beijing resident in the main city area will be 45 square metres, and in the suburbs will be 50 square metres.

However, while the green belt is increasing in the city area, there are two main problems. One is improper selection of plant species. Large areas of grass species, for example, which require high water consumption are planted in green belt areas in Beijing. The other problem is irrigation techniques used for the green belts. Poor management practice is the main cause of water being wasted during irrigation of the green belts. For example, irrigation should be by droplet or spray, but many places still apply wasteful hose irrigation methods. The low charge for irrigating water is another cause of wasted water in green belt irrigation in the city. Nowadays, the price of such irrigation water is lower than the civil water charge.

High water-consumption enterprises

In Beijing there are many commercial activities with high water consumption. Take, for example, vehicle washing. By the end of 2002, Beijing had 1.6 million vehicles, but this is likely to be at least 2 million nowadays. If, say, vehicles are washed once every five days then there are 240 000 cars, on average, which need washing every day. Water consumption is about 10–15 litres if using a recycling method. However, water consumption for washing a car by the traditional hose method needs 200–300 litres. The excess could therefore be as high as 69 600 cubic metres. There are 500 car-washing shops in Beijing and if the washing capacity of each shop is 200 cars a day, the city still needs 700 shops to meet the demand. However, there are many car-washing places in the suburbs of Beijing which use groundwater from wells without paying any costs, so there is no water-saving consciousness. Bathhouses are also high water-users. Beijing has 400 registered bathhouses, with an annual water consumption of 15 000 cubic metres each, which totals 6 million tonnes of water per year.

The production of clean drinking water is another high water-consumption business. At present, most of the enterprises are small companies with poor technical facilities. They produce clean drinking water but at a low rate of efficiency. One tonne of finished product uses three or four tonnes of water. However, the cost of producing such drinking water is excessively low: 40 yuan RMB for the use of tap water and 4 yuan RMB for the use of groundwater. Selling two buckets of product will recoup the cost of one tonne of groundwater, and one tonne of 'manufactured' water can fill 40 buckets. The industry does not seem to care about saving water, or about updating their facilities.

In 2004, the Beijing Municipal Government increased the price of water. Water for daily use is 5.4 yuan per tonne. Drinking water production companies and car-washing enterprises will pay nearly eight times that amount and bathhouse enterprises will pay more than 10 times. Excess water use will be charged several times more than the set rate. For example, if any bathhouse exceeds 40 per cent of the limit set, it will pay 61.5 yuan per tonne, plus 184.5 yuan extra. The Government hopes to control water wastage by imposing these fines.

3. Analysis of the water shortage situation in Beijing in 2010

Table 6.7   Available water supply in Beijing in 2010 (billion cubic metres)

Period	Normal year (50 per cent guarantee)	Slight low water year (75 per cent guarantee)	Low water year (95 per cent guarantee)	Available water	Scheme sources
Current status	4.420	3.724	3.120	4.068	Research Team on Water Resources and Urban and Rural Water Supply in Beijing 1997
2010	3.910	3.570	3.220	3.730	Natural Resources Series of China Editorial Board 1995
2010	4.544	4.211	3.856	4.366	Research Team on Water Resources and Urban and Rural Water Supply in Beijing 1997
2010	—	—	—	3.924	Research Team on Water Use in Beijing 1998

Source: Beijing Academy of Social Sciences (2001)

According to the Beijing Academy of Social Sciences, the available water supply for Beijing in 2010 will be about 4.366 billion cubic metres (Beijing Academy of Social Sciences 2001). This water supply includes recycled wastewater. Different water use schemes for Beijing in 2010 are listed in Table 6.8 (page 88). Compared to the relatively reasonable scheme mentioned above, the total water shortage in 2010 will be 0.534–1.359 billion cubic metres, daily water consumption shortage will probably reach 0.266 billion cubic metres, industrial water shortage will be 0.336–0.916 billion cubic metres, and agricultural water shortage will probably be 0.761 billion cubic metres. A detailed comparison is listed in Table 6.8.

Daily water use is relatively low in suburban and rural areas. In 1995, for example, the total daily water use in these areas was only 20 per cent of today's total daily water use. However, the next five years could see a rapid increase in

Table 6.8 Water use plan and estimation for Beijing in 2010 (billion cubic metres, per cent)

Industrial	Daily	Agricultural	Consumed in lakes and rivers	Total	Scheme sources
1.596 30.8%	1.335 25.8%	1.846 35.7%	0.400 7.7%	**5.176 100%**	Research Team on Water Use in Beijing 1998
1.950 35.9–34.0%	1.722–1.794 31.7–31.3%	1.462–1.681 26.9–29.4%	0.300 5.5–5.2%	**5.434–5.725 100%**	Research Team on Water Resources and Urban and Rural Water Supply in Beijing 1997
1.340 27.4–25.8%	1.072 21.9–20.6%	2.000–2.341 40.8–45.0%	0.450–0.500 9.2–9.6%	**4.900–5.200 100%**	Natural Resources Series of China Editorial Board 1995
1.004 23%	1.528 35%	1.528 35%	3.056 7%	**4.366 100%**	Beijing Academy of Social Sciences 2001

Source: Beijing Academy of Social Sciences (2001)

the amount of daily water use in suburbs and rural areas. In Beijing's general plan, daily water use in suburban/rural areas was set at 300 litres per capita per day. It is estimated that the total daily water use in suburban/rural areas will be 0.422 billion cubic metres by 2100, which will be 30 per cent of the total daily water use for the whole city. Therefore, the daily water use in the main city will be 1.106 billion cubic metres, according to an estimation made by the Beijing Academy of Social Sciences. In Beijing's general plan, daily water use in the main city area was set at 360 litres per capita per day. If the city uses 35 per cent of the total water supply, the average household water use for the residents in the main city area becomes only 128 litres per capita per day in 2010, which is lower than that of other megacities in China (Beijing Academy of Social Sciences 2001).

4. Some water-saving considerations

Water—the most common, important, useful and mysterious substance in the world—is no stranger to us. Water is all around us; it accompanies us all the time and serves us for industry and daily living. However, many of us pay little attention to water, for we waste it, pollute it and use it inefficiently without thinking of the consequences. This makes precious water resources even scarcer.

Beijing is a city where water resources are not naturally rich, but the demand for water is increasing with socioeconomic development and population growth. How to prevent water resources being a constraint to socioeconomic sustainable development in Beijing is a great challenge which needs our serious attention. What each citizen can do to reach the objective of sustainable water development is also an important issue needing careful consideration.

4.1 What are you willing to do?

- Increase community consciousness of saving water resources. People have always assumed that supplies of water are inexhaustible and have therefore used them extravagantly. Water resources in Beijing are not rich and tap water is hard-earned, because water is scarce per capita, is unevenly distributed geographically, and subject to unpredictable variations in a year, varying interannual change, and severe pollution. The significance of saving water is great, and the creed of 'Water-saving is glory and water-wasting is shameful' should be known by all.

- Form good habits in water-saving. There is evidence that up to 70 per cent of water could be saved if household bad habits were corrected. There are many habits that cause water wastage, some of which are only trivial, but they might include flushing away a cigarette stub or other small waste in the lavatory, running a tap for a long time just for a cup of water, peeling vegetables under a running tap, forgetting to shut off a tap (such as while opening a door for a guest, picking up the telephone or changing a TV channel), keeping water running when washing hands and face or brushing teeth, not checking a tap for leaks before sleep or an outing, and not repairing a leaking tap. Getting over these bad habits does not require much time or labour, but does have rewards.

- Use water-saving devices. There are many kinds of water-saving devices such as water-saving tanks, taps and toilets (which include the manual wrench and button types, or automatic models with electromagnetic induction and infrared remote control). Excess water consumption by older devices leads to extra cost.

- Check water leaks and water flow regularly. Check taps and pipelines at home and do not ignore leaks in a tap or pipe joint. Once a leak is found it must be repaired by either a plumber or by you as soon as possible. If it cannot be repaired immediately, close the general valve to prevent water being wasted.

4.2 How to save water

Washing machines

- When there are small numbers of clothes in a washing machine, a high water level will make the clothes wave around, and the lack of friction will not wash clothes clean. This also wastes water.

- Collect as many clothes as possible and wash them at the same time.
- Re-use the rinse water to wash the next washing load. This could save 30–40 litres of water.

Showers and baths

- Learn how to adjust the proportion of cold and hot water.
- Do not have the shower water running at maximum capacity.
- Do not wash your head, body and feet separately; try to get your entire body wet first, then use soap and rinse off.
- Keep your mind on bathing and make the best use of the time. Do not be carefree or chat while taking a shower, and do not play 'water battles' in the bathroom. Remember: time is water!
- Do not fill a bathtub right to the top; a third or a quarter of the bathtub capacity is enough.

Re-using water

- By washing the face first, then the feet, the water can then be used to flush the lavatory.
- Collect water in a big bucket, then use it to flush the lavatory.
- When cooking, the water used after washing rice or noodles could then wash utensils, as it removes oil easily and also saves water.
- In fish-breeding, the water used could irrigate flowers to accelerate their growth.

Washing utensils

- Use paper to wipe grease and food from dishes before washing them in hot water.
- Clean utensils with warm or cold water.

Playing games

- Do not play water-consuming games.
- Some toys (a water squirt gun, for example) use large amounts of water, which is not recommended, especially in places where water is scarce.
- Some children play water games beside water taps using hoses, splashing water in all directions. This wastes a lot of water.

Taking care of pipes

- Prevent water pipes from cracking when frozen. During winter, water pipes in Beijing are fragile, crack easily and can cause severe leakage. Special attention should be paid to regular prevention methods and to examining the pipes. For example, when surface earth is flushed away during the rainy season, the exposed pipes should be covered again before winter to prevent frost cracking the pipes.

- Water taps and pipes outside a house should be fitted with freeze-proof facilities, such as freeze-proof plugs and wooden boxes.
- Water pipes inside a house should be covered by pieces of sackcloth and tied by grass rope.
- Cracks around doors and windows should be sealed with paper to keep the house warm when there are water pipes inside the house.
- Frozen water pipes should not be baked by fire or heated with boiling water; the sudden expansion damages pipes and taps. They should be covered by a warm towel to help defrosting.

Recording water consumption
- There will be many benefits if you note down your water meter reading at a regular time every day, such as a fixed time in the morning or night.
- Water consumption every day, every month and every year can be easily checked, so you will be aware of the water charges that apply.
- If you do record your consumption, you will also know the amount of water you have saved. Whether there has been water wastage should be evident.
- Recording is easy for a day, but it can be difficult to persist for several months or years. With a little perseverance, stamina and patience, however, it can be achieved.
- Any change in water consumption will be obvious by recording, and you can note if the change might be caused by daytime temperature, cloudy weather conditions or whether is it raining.
- A change in water consumption might be caused by a change in living conditions, such as using an electric fan, or a new refrigerator or air conditioner, or a change in eating habits or food.

4.3 Things that you can do

- Measure the flux of tap water. Flux is the quantity of water discharge (flow past) at a given unit of time (for example, a second). The principle of the measurement of flow rate in a house is the same as in rivers, but the method is different. Prepare a bucket, calculate its volume (18.6 litres, for example) and then get a stopwatch or use a wristwatch with a second-hand sweep. Open the water tap as widely as it will go for full volume, fill the bucket, then read out the time it has taken to fill the bucket (23.4 seconds, for example). The measured flow rate is the volume of the bucket divided by the time taken to fill it, i.e. $18.6/23.4 = 0.795$ l.s^{-1}. You should measure three times and use the mean result.
- Experiment with water-saving. It is interesting to do a simple experiment when you come across a water-saving method. For example, you can calculate the change of water consumption by measuring the flow rate when a brick is

placed in a lavatory water tank. It will be easy to estimate the change of water consumption in a day or a month from the record.

- Try to be a water-saving pioneer. Saving water is the responsibility of each person. Only when we all begin to pay attention to water-saving can water shortage be eliminated. Our lifestyles can then be stable and in harmony with the environment, and our living environment can be made beautiful and comfortable. The younger generation should be actively involved in water-saving, and they need to encourage friends and relatives to do the same. This contributes to our society.

5. Water-saving cases in Beijing

5.1 Schools

Beijing No. 4 High School

The principles of water management in Beijing No. 4 High School involved replacing outdated water equipment, setting up a water use management system, and expanding water recycling and re-use. The school made a large investment to improve equipment by using infrared, automatic on–off remote controls in bathrooms to stop the flow of water automatically after people left. Induction-type flushers and foot-pedal flushers in toilets were also installed. The school also established a water-saving management system, a maintenance system and a water consumption quota control system, then distributed the quota to every water consumption part, and assigned a person to check water meters to find whether there were any unusual water use conditions. Also, the lawn in the school was irrigated using micro-spray systems. These improvements have saved about 2000–3000 cubic metres of water each year.

Beijing Jiaotong University

Beijing Jiaotong University established water-saving management systems, used reclaimed water, installed water-saving equipment and promoted water-saving education. Specific management measures included formulating a water consumption quota control system, using a regular patrol to check systems, and set up a reward or penalty system. Green land was irrigated using reclaimed water and micro-spray technology. The university also installed water-saving equipment, especially the new press type of tap in flushing sinks in the school dining room and dormitory. By separating rainfall and drainage pipelines, the collected rainfall was recycled for green belt areas. Posters and slogans were also widely used as part of the education campaign to save water. The university saves up to 300 000 cubic metres of water each year with these measures.

5.2 Parks

The Wanshou Park is one of the Guanyu Temples built in the Ming Dynasty, the year of Wanli-45 (1617). Its former name was Wanshou West Palace because it is located west of the Hongren Wanshou Palace (destroyed in the Qing Dynasty). The park is an entertainment and exercise area for nearby residents. In 1995 the Garden Management Bureau of Xicheng District of Beijing invested 13.8 million yuan to comprehensively rebuild this desolate and unkempt place. Wanshou Park is the first to be specially designed for aged people in Beijing. The park has a total area of 47 000 square metres, with 30 000 square metres of green belt and more than 10 000 trees. The green cover rate is 79.27 per cent.

The Wanshou Park management took on the general requirement of the municipal government by 'taking water saving as an important strategy in sustainable economic society development for the capital city'. The park authorities recently built a rain recycling irrigation system. The system joined rainfall pipelines to a filter well, collected surface run-off, allowed rainfall from gutters in the green belt to be re-used, increased water infiltration to underground water in the flood season and saved water in lawn irrigation. The park once used 36 000 cubic metres of water every year to maintain green land, but now the park saves 10 000 cubic metres of water through these measures. The park has made investments in irrigation equipment, and since 2003 has replaced more than 70 sets of sprinkler irrigation facilities, gates, water pipelines and so on. The new spray irrigation facilities avoid inundating the green belt with excess water. Communal toilets in the park have also been updated by using water-saving fittings.

5.3 Hotels

There are 106 hotels receiving foreigners in the urban planning area of Beijing, comprising 16 five-star hotels, 32 four-star hotels and 58 three-star hotels. In total, these hotels use 32.9 million cubic metres of water annually, which is 3.69 per cent of the total water consumption in the urban planning area. Therefore it is extremely important that water saving consciousness is strengthened and that there is an increased investment in water-saving technology by hotels.

Peace Hotel
Water-saving strategies for the Peace Hotel include establishing a water-saving management system, strengthening appliance management, and using water-saving propaganda for education. The hotel:

- assigned special energy conservation supervisors to monitor and control all water equipment in the hotel, who patrol and check water use regularly every day

- transformed the cooling tower. Electronic water treatment equipment was installed, which means that there are no chemical pollutants in the sewage. Therefore this water can be used in staff restrooms, which can save up to 300 cubic metres of water each month
- transformed dry-washing machines, fixed cooling recycle equipment, and recycled water in the hotel
- replaced regular water fittings with water-saving products, especially in water tanks in bathrooms
- introduced education campaigns on water use through posters, the in-house news-paper 'Sound of Peace' and through the hotel's Peace TV channel. Competitions based on water-saving knowledge have also been organised.

Figure 6.5 shows the water savings at the hotel.

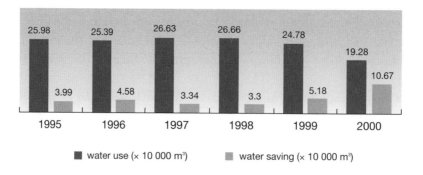

Figure 6.5   Water saving effect on the Peace Hotel, 1995–2000

Source: Beijing Water Saving Management Office (2005)

Hong Kong & Macao Centre

Water-saving strategies at the Hong Kong & Macao Centre mainly include intro-ducing advanced management experience, strengthening water saving manage-ment and utilising reclaimed water. Table 6.9 shows an example for daily resources consumption.

The main strategies include:

- formulating a water-saving management system
- learning from advanced management experience, such as formulating reasonable water consumption parameters for various departments and auditing water consumption each month. If a department is found to be using more than the standard water quota, the cause of the water consumption in the department can be determined
- patrolling areas of the Centre at least four times a day to check the water use, and setting up a sign-in book at each place

Table 6.9   Re-use of reclaimed water at the Hong Kong & Macao Centre

An example in April 2001							
Item	Daily value	Actual consumption yesterday	Daily difference	Sum of daily value	Sum of actual consumption	Sum of difference in current month	Sum of difference month by month
Electricity (kWh)	21 000	19 200	–1800	1 798 000	1 828 866	–8560	30 866
Water (m³)	633	628	–5	53 475	45 276	–38	–8199
Oil (litres)	1143	950	–193	100 182	84 100	–679	–16 082
Natural gas (m³)	460	280	–180	44 944	24 220	–600	–20 724

Source: Beijing Water Saving Management Office (2005)

- establishing a fast response time for maintenance: once a leak has been detected, the repair should be done within 5–7 minutes after receiving the repair request
- strengthening routine maintenance of reclaimed water equipment to guarantee its normal operation.

Figure 6.6 shows how reclaimed water is re-used in the hotel each day. This is a total of 300 cubic metres per day, and 10 950 cubic metres per year.

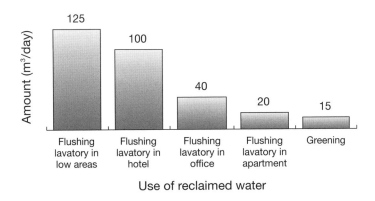

Figure 6.6   Re-use of reclaimed water per day at Hong Kong & Macao Centre

Source: Beijing Water Saving Management Office (2005)

5.4 Factories

The Furniture Branch Company of the Beijing Tiantan Limited Company Tiantan has two production lines to spray powder on metal furniture, which is the main use of water in this company. In order to solve the problem of highly concentrated wastewater, the company set up a project of wastewater treatment and re-use. This began in August 2003, and was completed in March 2004. Treated wastewater is measured up to the national surface water discharge standard, which satisfies primary clean and pickling water recycling in the company, and also can be used in green belts. This project has a wastewater treatment capacity of 45 000 tonnes per year and has disposed of 35 000 tonnes per year in recent years.

There are two kinds of wastewater: one is pickling wastewater, and the other is electrophoresis wastewater. The pickling wastewater is first neutralised with an alkali to a certain pH value, then ferric oxide $Fe(OH)_2$ and $Fe(OH)_3$ is deposited. This absorbs pollutants such as suspension substances (SS) and chemical oxygen demand (COD) in wastewater in the deposition process, and the pollutants are separated using magnetic separators to purify wastewater. For the electrophoresis wastewater, 'medicine' is added to coagulate and deposit, then the major pollutants such as COD, biochemical oxygen demand (BOD) and SS in wastewater are eliminated, and the remains are disposed of by biochemical processing. Finally, activated carbon is used to filter and degrade organisms, certain ions and the peculiar smell, thus guaranteeing that the electrophoresis wastewater can achieve surface water discharge standards.

Under pressure, the clean water from the above two treatment processes are passed through two levels of sand leach, and are finally re-used for pickling water and primary clean recycled water, which together reach 35 000 tonnes per year. The green belts in the factory area also use the clean water purified in this project. Consequently water utilisation efficiency has been improved enormously.

6. Objectives for constructing a water-saving society in Beijing

In order to promote water-saving in a universal way, work on the program to establish a water-saving society in Beijing began in 2005. To create a water-saving society, the value of saving water needs to be recognised as part of society's ideology, and perceptions of water-saving need to be transformed to foster good water-using habits and encourage scientific developments to enhance water use.

This program of establishing Beijing as a water-saving society aims to:

- plan economic and social water demands
- plan harmonious development between humans and nature, between urban and rural areas, and among regions as a whole
- make possible the rational allocation of water resources (rescannable extraction and recharge) and water recycling

- provide guarantees of enough water resources in order to realise the objectives of 'New Beijing, New Olympic Games'
- capitalise on sustainable socioeconomic development.

By the year 2010, it is hoped that Beijing will be established as a water-saving society, and will achieve:

- the improvement of the water-saving consciousness of all citizens
- the establishment of the long-term efficiency of water resource management mechanisms to utilise water more efficiently
- the setting of reasonable water quotas

Table 6.10   Specific control index for Beijing to establish a water-saving society by 2010

Items	Specific control index
Gross fresh water	Low flow year: 3.36 billion cubic metres Normal flow year: 3.65 billion cubic metres
Recycling water use annually	0.46 billion cubic metres
Water withdrawal quota per 10 000 yuan GDP	52 cubic metres
House water consumption per urban capita	120 litres/person/day
Water withdrawal quota per 10 000 yuan increase in the tertiary industry	17 cubic metres
Water withdrawal quota per 10 000 yuan increase in industry	33 cubic metres
Water consumption per 10 000 yuan increase in agriculture (including recycling water)	1087 cubic metres
Fresh water consumption per 10 000 yuan increase in agriculture	730 cubic metres
Water re-use rate in industry	≥ 93%
Water consumption of single main product	Achieve the domestic highest level and medium developed country's advanced level in that very year
Popular rate of water-saving appliances in town	100%
Agricultural water-saving irrigation rate	100%
Urban and rural effluent treatment rate	90%
Industrial water use and agricultural irrigation water	Zero growth
Public living water use efficiency	Reach international advanced level of that time

Source: Beijing Water Saving Management Office (2005)

- the strengthening of management to realise the macro allocation of gross water consumption
- the full application of market mechanisms to increase water-saving investments and finances, and the adjustment of water consumption by price
- the development of industries which conform to the character of the capital city
- the establishment of a sound water-saving standard system, and a water-saving scientific and technological support system, to standardise water use management
- legislation to perfect the water-saving laws and regulations
- an improved propaganda function of public media to set up and perfect social participation for water-saving.

Establishing a water-saving society in Beijing will occur in three stages.

- In 2005–06, the primary stage, the essential target is to ensure urban and rural water supply when confronted with water scarcity.
- In 2007–08, the advanced stage, the main object is to guarantee water resources for 'New Beijing, New Olympic Games'.
- In 2009–10, the final stage, Beijing will become a standard water-saving society (Table 6.10, page 97).

Acknowledgment: The authors extend their thanks to the Beijing Water Saving Management Office for help in completing the report and providing data.

Bibliography

Beijing Academy of Social Sciences 2001, 'Population capacity model and forecast under the constraint of water resources in Beijing', *Social Sciences of Beijing*, vol.1, pp. 35–8.

Beijing Water Saving Management Office 2002, 'Countermeasures for realization of sustainable water resources use in Beijing', *Energy conservation and environmental protection*, 03–25–27.

Beijing Water Saving Management Office 2005, *Report of water saving in Beijing*, Beijing Water Saving Management Office, Beijing.

Bureau of Beijing Hydrology 2004, *Beijing Water Resources Bulletin, 1996–2003*, Bureau of Beijing Hydrology, Beijing.

Han, G & Wang, L 2000, 'Strategies of water resources in Beijing in the new era', *Journal of Peking University (Philosophy and Social Sciences)*, vol. 37, no. 6, pp. 118–26.

Hui, S, Xie S & Zhang, S 2001, 'Optimizing distribution of water resources in Beijing after water is available from south-to-north water transfer project', *China Water Resources*, vol. 1, pp. 14–15.

Li, X & Zhang, W 1996, 'The optimization of water resources and industrial structure in Beijing', *Earth Sciences Frontiers*, vol. 3, nos 1–2, pp. 249–54.

Qian, Y 2001, 'Review on water resources protection strategy in Beijing', *Beijing Water Resources*, vol. 6, pp. 1–3.

Wan, Y & Jin, D 2001, 'Emergency countermeasures on water resources in Beijing', *China Water Resources*, vol. 7, pp. 56–7.

Wang, R 2003, 'Reasonable allocation of water resources in Beijing at the beginning of 21st century', *Beijing Water Resources*, vol. 2, pp. 33–4.

Yan, C 2000, 'Protection and reasonable utilization of water resources in Beijing', *Beijing Planning Review*, vol. 6, pp. 26–9.

Zheng, G & Lu, J 2001, 'Water resources in Beijing', *Chinese Geology*, vol. 28, no. 4, pp. 45–8.

Zhu, Z & Shao, Z 2003, 'Grim situation of water resources in Beijing', *Capital Economy*, vol. 7, pp. 27–8.

7 | Rainwater harvesting in Mumbai

Shyam Asolekar

1. Introduction

India, although one of the wettest countries in the world, has been beset with severe water scarcity. There is great pressure on water resources throughout the continent owing to an increased population in cities, high industrial processing needs for water, and the need for water for irrigation, and this is further complicated by industrial pollution and agricultural run-off. India stands at the crossroads where sustainable consumption of resources and increasing industrial and agricultural activities are diverging. It is becoming increasingly clear that if we fail to adopt and implement institutional policy and technological reforms to address both the shortage and the declining quality of water supplies, water will limit growth in India's cities and farms.

The predicament of the megacity of Mumbai is even worse. Being the financial capital of India, Mumbai attracts people from across the country and thereby faces ever-increasing water demand. The Municipal Corporation of Mumbai brings drinking water from several sources situated up to 100 kilometres from the island of Mumbai. It is becoming not only more expensive each year but politically it is becoming unviable. In spite of all the efforts made by the city administration, nearly 50 per cent of the population of Mumbai (nearly seven million people) do not have an adequate water supply. The peninsula of Mumbai and the surrounding region receives plenty of rainwater every monsoon (nearly 2500 millimetres, or 100 inches), but in every monsoon season, which lasts from mid-June to the end of August every year, a large quantity of water simply runs off after each rain shower. It makes sense to harvest even a fraction of this water.

This case study addresses the critical issues related to water management in a typical urban community, especially in the megacity of Mumbai. There are four parts.

- It delineates the significance of rainwater harvesting (RWH) in the context of Mumbai's water scarcity, followed by an articulation of the role of small streams and headwaters in the river ecosystem. An attempt has also been made to provide information on the status of Mumbai's urban aquatic ecosystem.

- It attempts to depict the needs and implications of action-learning projects taken up by graduate students at the Indian Institute of Technology, Bombay, and their respective findings including the constituents of harvested rain and treatment options, traditional rural and contemporary rainwater-harvesting (RWH) systems, case examples from rural and urban India, legislation which encourages RWH in India, and the application of remote-sensing data and GIS in RWH.
- It describes findings of the skill-building workshop on RWH.
- Appendices 7.1–7.2 provide information on water budgeting, and on the design of RWH projects for schools and homes which could be useful in action-learning projects in high school and junior colleges.

2. Significance of RWH in the context of Mumbai's water scarcity

Mumbai is governed by local self-government known as the Municipal Corporation of Greater Mumbai (MCGM) and consists of the island city of Mumbai and its suburbs. In the twentieth century, the city was known as Bombay. The area under the jurisdiction of the MCGM is approximately 438 square kilometres. In the last 50–60 years, Mumbai has changed in terms of population, housing, traffic density, land use, environment and public health. The population has increased from six million in 1950 to approximately 18 million today. The city also supports another two million as a 'floating' population, commuting to the town and suburbs every day for business and work from nearby cities and communities.

Although the water supply in Mumbai is considered as one of the most reliable and advanced systems in India, water scarcity in Mumbai is not much different from other parts of the country. At the time of independence in 1947, Mumbai was supplied with 494 million litres per day (where one million litres = 1000 cubic metres) of water from the city's own water bodies at Tulsi, Vihar and Tansa Lakes, and it was sufficient for the city's needs. Presently, the city water supply is augmented with external water resources using Vaitarna, Upper Vaitarna and Bhatsa reservoirs (created by damming the rivers) and so the MCGM has increased its water supply to 2900 million litres per day. Unfortunately, this amount of water is not sufficient for the city of Mumbai since overall demand is actually 3400 million litres per day. The shortfall of 500 million litres per day is expected to rise to about 2000 million litres per day by 2021 (MCGM 2003).

The residents of Mumbai and the elected representatives in the MCGM regularly debate on potential avenues for obtaining additional potable water to satisfy the ever-increasing demand. It appears that there is no other option but to look for water management interventions including conservation, groundwater pumping and the re-use and recycling of treated sewage as the most promising solution for RWH.

In reality, every drop of sweet water on the earth is harvested from rain. For example, water reservoirs created by dams, artificial ponds and lakes get filled with rainwater through the hydrological processes in their respective watersheds. In an urban context, however, RWH refers to the act of deliberately harvesting rainwater from rooftops and paved surfaces. The collected rainwater can be suitably treated and used in a variety of applications including processed water for industries, domestic grey water applications, drinking water supplements, or flushing and gardening.

In many instances, it is not economic to store harvested water for long periods because of the cost associated with the construction and maintenance of huge tanks, so the harvested rainwater is simply recharged into the ground. The following section elaborates on the potential ecological benefits and the sheer necessity of such recharging in the Mumbai peninsula.

3. The role of small streams and headwaters in river ecosystems

It is well known that the inhabitants of the watershed receive a variety of benefits from the small streams and headwaters (the smallest streams in the network), as well as wetlands. These benefits include flood control, adequate high-quality water, and habitats for a variety of plants and animals. The watershed offers inherent special physical and biological characteristics—groundwater recharge, trapping of sediments and pollution from fertilisers, nutrient recycling, creation and maintenance of biological diversity, and sustenance of the biological productivity of downstream rivers, lakes and estuaries. The hydrological, geological and biological characteristics of small streams and wetlands require protection because healthy headwater systems are critical to the healthy functioning of downstream streams, rivers, lakes and estuaries (Meyer et al. 2003).

In the context of Mumbai, the nexus between trees, vegetation, green cover, soil erosion, damming of monsoon flood waters, and siltation in lakes and ponds has been recognised in recent times. It has been understood that all uplands play a crucial role in controlling run-off in all the river/nalla systems and in the lakes and ponds in Mumbai. If Mumbai's people are to enjoy a respectable quality of life, we must move quickly toward a sustainable future, which involves managing these aquatic systems with the principles of sustainable development in view, and encouraging wide participation through partnerships and networked institutions. Key to sustainable development is the empowerment of the residents through action-oriented partnerships at all levels.

The rejuvenation and environmental upgrading of hills, slopes and lakes/ponds in the Mumbai region has been undertaken to minimise topsoil erosion, enhance groundwater recharging, improve flows in rivers, and improve the ecosystem in several other ways. However, the environmental upgrade should include measures which will eventually prevent any interference to the existing small streams, headwaters and wetlands of Mumbai. Fortunately, the municipal authorities have begun to encourage advance locality management (ALM) groups in maintenance and vigilance, and to consider public–private partnership (PPP) for bringing in capital and professional management.

Thus, it should be remembered that the green belts, gardens, hills, sandy beaches, and the sediments in creeks, estuaries, rivers, lakes and oceans perform a crucial ecological function in a given ecosystem. Mumbai's prevailing challenges— population pressure, frequent occurrence of flash floods and increasing possibility of pollution of water bodies—can only be addressed through the rejuvenation and environmental upgrade of the urban ecosystem, which will prove to be of even more significance in the near future.

4. Status of Mumbai's urban aquatic systems

Ecosystems are made up of several subsystems and those subsystems are further made up of sub-subsystems. A significant impact on any of the subsystems or even on sub-subsystems can eventually have repercussions for the entire ecosystem. The impacts leading to disturbance of a typical urban ecosystem like Mumbai can broadly be divided into two categories: natural calamities (which mostly tend to pose sudden or acute disturbance to the system), and inappropriate and inadequate arrangements in the management systems. See Figure 7.1.

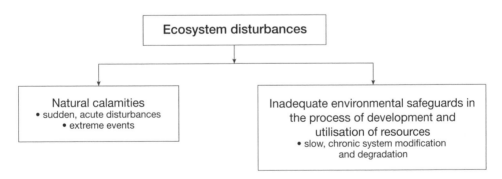

Figure 7.1 Ecosystem disturbances

Excessive rainy-season flooding and the year-long extreme deterioration of the quality of water flow in the stream channels have been the principal adverse impacts on the ecosystem of the last 150 years of development in Mumbai. Once Mumbai was a cluster of seven independent islands. Today this is a reclaimed and altered terrain. The MCGM area appears to have progressively drifted away from its natural state. As a result, the natural processes of silting, erosion, estuary movement and the interaction of mangroves with tides have been greatly affected. All the river ecosystems in Mumbai have either completely drifted away or have been drifting away from their so-called natural state because of land reclamation activities and the developments on those lands.

For example, the 1944 map of the *Survey of India* does not mention the Mithi River, but shows the Mahim River with a tidal range up to Bail Bazar (Old Kurla). Even the mouth of the river was then unclear because of the mudflats along and between Bandra, Dharavi, Sion and Kurla. The river channel in a recognisable form appears to have developed subsequently, when mudflats were occupied through the raising of ground levels for reclamation. The area between Salsette Island and Mahim has also undergone considerable topographical modification.

Very few patches of Mumbai's ecosystem today are in fact survivors of the pristine 150-year-old ecosystem. The area west of the Manori Creek is, however, one example. Some patches, however, are present in today's modified ecological form because they adapted to human intervention and forces of development. These include Mahul or places near Manori Creek. Superimposed on the earlier natural ecosystem in and around the islands of Mumbai are artificial civil constructions.

It has been recognised that Mumbai needs special consideration in its management, because it is a coastal island-city with an excessively high population density. While responding to these challenges of development, Mumbai Island has undergone a series of reclamations and transformations. Mere restoration of watercourses and water bodies in Mumbai will not provide the needed flood protection. There is certainly a need for much higher carrying capacity of floodwater and wastewater in the rivers and creeks. However, the inflow itself needs to be regulated through environmental management so that future floods can be handled via appropriate engineering interventions in conjunction with improvement of the waste processing ability in the riparian zones, riverbeds and creeks.

Mumbai's urban ecosystem is made up of ten subsystems including four rivers, Powai Lake, Sanjay Gandhi National Park (including Tulsi and Vihar Lakes), creeks, bays, the Mithi River estuary, and the coastal zones. Each of these four river basins comprises several small catchments. Encroachments in the riverbeds or on the banks of the rivers in Mumbai have choked and pinched the watercourses and aggravated the risks of flooding. Urban activities responsible for environmental degradation include encroachment leading to a narrowing of banks, reclamation of riverbeds for housing, unauthorised slum development, the construction of industrial units, diversion of the river flow due to dumping of construction debris and solid wastes, and wastes generated by stables and cottage industries on the banks of the rivers.

Due to developments in the Mumbai region, almost all natural areas have been covered with asphalt or cement roads, with pavements for parking, and concrete buildings. As a result of this development, the rainwater run-off is now almost 100 per cent, so water percolates into the ground and all the natural processes which could recharge groundwater have been dismantled.

In addition, several housing complexes and industries are pumping out groundwater for use in industrial processing and greywater applications. In Mumbai and in almost every coastal city in India, there is the threat of groundwater intrusion and salinisation of sweet water aquifers. Metropolitan cities like Chennai have already salinated their sweet water aquifers due to the intrusion of seawater. The current status of the water crisis and the situation ahead has not been brought about by natural factors like drought. It is due to ever-increasing population pressure, urbanisation and industrialisation.

The situation of midland metro-cities is equally worrisome. Excessive pumping of underground water has depleted and depressed the water table to alarming depths in large metro-cities including Delhi and Ahmedabad. In all such circumstances, RWH seems to be a panacea for revitalisation of groundwater levels and quality. On the Mumbai peninsula, more specifically, RWH appears to be the only solution to recharge groundwater aquifers beneath the city.

5. Constituents of harvested rain and treatment options

Typically, constituents of harvested rainwater will depend on two factors: the ambient air quality, and the cleanliness of the collection spots (roof or ground). The constituents could be microorganisms, totally dissolved solids (mostly in the range of 1–50 parts per million) and totally suspended solids. The water could also be acidic if the area is prone to acid rain (Undegaonkar 2005).

The treatment strategy for harvested rainwater would depend on its end use. For example, where the end use is, say, gardening or floor-washing, the rainwater should only be subjected to screening, followed by settling. Much more treatment will be required if the harvested water is to be used as drinking water. In general, the following steps are suggested for the treatment of harvested rainwater.

- **Screening**. In this step the larger particles are removed. Screening prevents leaves and other debris from entering the storage tank.
- **Settling**. In this step, settling tanks are used to remove silt and other floating impurities from rainwater. A settling tank is like an ordinary storage container having provisions for inflow (bringing water from the catchments), outflow (carrying water to the storage tank through a filtration unit) and overflow. A settling tank can have an unpaved bottom surface to allow standing water to percolate into the soil. Any container (masonry or concrete underground tanks, old unused tanks, pre-fabricated PVC or ferro cement tanks) with adequate storage capacity can be used as a settlement tank.

- **Filtration**. Sand and cloth filters are found to be useful for this purpose. The filter is used to remove suspended pollutants from rainwater. A filter unit is a chamber filled with filtering media such as fibre, coarse sand and gravel layers to remove debris and dirt from water before it enters the storage tank or recharge structures. Charcoal can be added for additional filtration. Charcoal and sand filters are commonly used.
- **Disinfection**. This might include the use of ultraviolet light, or chlorine, which is used commonly because of its easy availability, low cost and high disinfection efficiency.

Any level of society can apply the abovementioned methodologies for the treatment of rainwater, as they are very easy to understand. The required components of the treatment system are easily available and are also economical.

If harvested rainwater is to be used as drinking water, however, it needs advanced tertiary treatment, including processes such as reverse osmosis, ultrafiltration and ozonation.

6. Traditional rural and contemporary rainwater harvesting systems

RWH has always been practised in India, especially in areas which are prone to drought. There were several methods which were traditionally practised in different parts of the country. Traditional RWH, which is still prevalent in rural areas, uses surface storage bodies like lakes, ponds, irrigation tanks, temple tanks and so on. Bara (2005) has reviewed the literature available on various methods of water harvesting systems used traditionally in the past and are still used in rural areas. Some of them are described below.

6.1 Kunds

Kund is the local name given to a covered underground tank and was developed primarily for the storage of drinking water. The kunds are made of local materials or cement and are more prevalent in the western arid regions of Rajasthan, where the limited available groundwater is moderately to highly saline. The first known construction of a kund in western Rajasthan was in 1607.

6.2 Tankas

Like kunds, the *tankas* (or small tank) is a kind of underground tank found traditionally in most Bikaner houses, and is especially meant for harvesting of rainwater for drinking. They are built either in the main house or in the courtyard and are structured like circular holes in the ground, lined with fine polished lime, in which rainwater is collected. Tankas are often beautifully decorated with tiles, which help to keep the water cool.

6.3 Kere

Tanks, called *kere* in Kannada, are the predominant traditional method of irrigation in the Central Karnataka Plateau. In this system, channels branch off from anicuts (or check dams) built across streams, or by streams in valleys. The tanks are built in series, usually situated a few kilometres apart; the outflow of one tank supplies the next, all the way down the course of the stream. This ensures that there is no wastage through overflow, and the seepage of a tank higher up in the series is collected in the next lower one.

6.4 Vav, vavadi, baoli and bavadi

Traditionally, stepwells are known as *vav* or *vavadi* in Gujarat, or *baolis* or *bavadis* in Rajasthan and northern India. Stepwells are used for collecting water for drinking purposes. When stepwells are used exclusively for irrigation, a sluice is constructed at the rim to receive the raised water and lead it to a trough or pond, where it runs through a drainage system and is channelled into the fields. A major reason for the breakdown of this traditional system is the pressure of centralisation and agricultural intensification.

6.5 Zings

Zings are another kind of water-harvesting structure found in ancient India at Ladakh. They are small tanks which collect water from melted glaciers. In this system a network of guiding channels brings the water from the glacier to the tank. As glaciers melt during the day, the channels fill up with a trickle that in the afternoon turns into flowing water. The water is collected and stored for later use.

6.6 Chaukas

This system is used in the village of Laporiya at Dudu Block, Jaipur, in the state of Rajasthan, where dyked degraded pastures are used to harvest rain. *Chaukas* are rectangular plots in a dyked pasture which store rainwater. They are 66 metres long and 132 metres wide and the enclosures are arranged in a zigzag pattern and lie along a small gradient. Dykes 1.5 metres high are built along the three sides that lie towards the lower part of the land. Trees are planted on these dykes to give them additional support to withstand rain.

When it rains, water collects in the dyked lower half of the chauka. As the amount of water stored in the enclosure rises, it flows into the neighbouring chauka, and so on, gradually seeping over the entire pasture. This means that fields are never inundated with water. After reaching the last chauka, the water flows into a monsoon drain. This system not only provides adequate water for villagers, but also promotes the recharge of groundwater.

7. Rainwater harvesting cases from rural India

7.1 Bhaonta-Kolyala (Johads)

The residents of Bhaonta-Kolyala were facing the problem of acute scarcity of water. A local non-government office called Tarun Bharat Sangh made them aware of the importance of *johads* for their village. A johad is a crescent-shaped earthen check dam, which is also considered a divine place by village inhabitants. They visit the johad on all the happy occasions of life, such as births and marriages, to be blessed by the deity residing in the johad.

The residents of Bhaonta-Kolyala have revived the johads, hoping that they will help them to overcome the problem of water scarcity. They began protecting forests and repairing old johads. They mapped the natural drainage system and chose tentative sites to construct new johads. The aim of the villagers is to catch all the rainwater falling on the village. In 1988, repair work on the johad started. When the monsoons arrived, the johad was filled with water. Overwhelmed by the results from a single johad, the villagers started building more structures. Today, the village has a total of 15 water-harvesting structures, including a concrete dam 244 metres long and 7 metres tall in the upper catchments of the Aravalli to stop water before it flows downstream. Construction started in 1990. The dam has been a turning point for the villagers and it has helped to curb migration. Instead, villagers are returning. After just one year the water level in the wells downstream rose by two to three feet. Today, all the agricultural land is under cultivation. Every rupee invested in a johad has increased the village's annual income by a factor of three.

7.2 Balisana

In Balisana village, in the Patan district of Gujarat, a long canal of 3.05 kilometres was desilted and rainwater was diverted and collected in a recharge well. The cost was 52 lakh, out of which the government contributed 60 per cent and the remaining 40 per cent came from *shramdaan* (a temporary workforce) by the villagers. The rainwater harvesting has filled the old well with fluoride-free drinking water.

7.3 Darewadi

In this village, in the Ahmednagar district of the state of Maharashtra, there are 197.23 hectares of seasonally irrigated land and 737.62 hectares of rain-fed irrigated land. This area was facing drought for many years and it affected the agricultural productivity of the village. The villagers decided to harvest rainwater and they invested Rs 17 000 and constructed a road. The Watershed Organization Trust, an Ahmednagar-based non-government organisation, helped them to harvest the rainwater. They constructed contours, trenches, gullies, plugs, farm bunds and check dams. The program was started in 1996 and within two years the positive changes could be seen. Agricultural productivity increased, and the surplus was sold in the market. In addition, 1500 litres of milk was produced, which was sold in the cooperatives.

7.4 Sukhomajri

This village is in the Haryana area and faced a severe water crisis. In 1977, four tanks were constructed to increase storage capacity. The result was a vast improvement—the crop yield rate increased from 6.83 quintal per hectare in 1977 to 14.32 per hectare in 1986. The annual household income rose to Rs 3000.

7.5 Ralegaon Siddhi

This village is situated in the Ahmednagar district of Maharashtra. The total cultivable land available was 1700 acres (out of a total of 2200 acres). Annual rainfall in this area averages 370–400 millimetres. Before the villagers began actively harvesting rainwater, they had been facing the problem of drought for several years. The land is undulating but the soil quality is poor and this leads to rapid run-off of rainwater. To capture that rainwater the villagers repaired wells and constructed percolated tanks. Rainwater harvesting brought fruitful results. Agricultural water increased from 100 hectares to 1500 hectares. Per capita income has risen from Rs 271 in 1975 to 2200 at the present time. About 40 per cent of households have an annual income of more than Rs 48 000.

8. Rainwater harvesting cases from urban India

As stated earlier, the urban water crisis could be solved to a certain extent by using rainwater harvesting techniques. Today, the RWH systems are used on a large scale in cities like Chennai, Bangalore and Delhi where RWH is a part of state policy. Tamilnadu is the only state to achieve 100 per cent coverage in rainwater harvesting in all government buildings, private buildings, institutions and commercial buildings in all the 30 districts of the state. Elsewhere, countries like Germany, Japan, the United States and Singapore are also adopting RWH systems.

The Centre for Science and Environment (CSE), a non-government office based in Delhi, completed a study on the usefulness of RWH for Delhi. Data from 11 rainwater-harvesting projects spread across Delhi showed an increase of 5 to 10 metres in the groundwater levels over two years. On the basis of the study, the CSE made some suggestions to enhance rainwater harvesting in urban areas.

- Increase the harvesting of rainwater at the city level by protecting and regenerating the tanks and ponds.
- Promote individual rainwater harvesting projects at the colony and household level, which improves the local groundwater levels.
- Ensure by law that all commercial users are required to harvest rainwater.
- Redistribute the water supply across the city to ensure equity. The availability of water in the city is adequate (over 200 litres per capita) but the problem is that while certain areas (New Delhi and the Cantonment) get over 400–500 litres per person, other areas do not get even 30 litres per person.
- Treat all the sewage collected in Delhi and recycle the water so that there is no pollution of rivers to destroy the water cycle.

According to Sekhar Raghavan, a RWH expert from Chennai, there are three components which are crucial in promoting urban RWH:

1   Education: Creating an awareness regarding the importance of rainwater harvesting, both for immediate use and also for sustaining the water table in the long run.

2   Implementation: Designing and providing programs to help citizens implement rainwater harvesting in an efficient and cost-effective manner.

3   Evaluation/Research: Carrying out studies on:
    - the nature of the sub-soil in different city neighbourhoods and its capacity to absorb large quantities of injected rainwater;
    - the effectiveness and adequacy of various types of RWH structures; and
    - the post-monsoon impacts on the quality and exploitable quantity of groundwater in places where RWH has been implemented. (Raghavan 2004)

To date, several cities around the world are adopting, or are in the process of adopting, RWH systems as an appropriate water management tool. Some case studies from urban India are shown below.

8.1 Panchsheel Colony, Delhi

There is a small housing community called Panchsheel Colony in Delhi, and in that area the water level was declining rapidly. To solve the water problems, the Resident Welfare Association established a rainwater harvesting system for the entire colony. The water supply is mainly from six bore wells. A network of stormwater drains in the entire residential area is used for harvesting rooftop rainwater and surface run-off. About 36 recharge wells measuring $1 \times 1 \times 2$ metres were constructed in the stormwater drain for facilitating groundwater recharge. The water passes through a bed of pebbles and is collected in the wells.

8.2 Tihar Jail, Delhi

In this case, the barracks of the Tihar Jail were flooded for hours after every rainfall, but the water level in the jail was 14 metres below ground level in April 2003. The Tihar Jail authorities therefore decided to adopt an RWH system. The total area available for the harvesting of the rainwater was 4125 square metres and the total volume of water they could harvest was 1280 cubic metres, or 1.28 million litres.

8.3 Chennai, State of Tamilnadu

In Chennai's coastal suburbs, seawater has intruded into the coastal aquifers, rendering groundwater quite saline. An inadequate supply of water over the last two decades has forced the general public to relentlessly exploit the groundwater, which in turn has depleted the quantity of groundwater and also deteriorated the quality of groundwater.

To solve the problem, RWH was initiated. In two of Chennai's coastal suburbs, Besant Nagar and Valmiki Nagar, door-to-door campaigning was started to create awareness about the system. It was explained to the residents that implementing RWH would be simple and cost-effective. Many non-government organisations, such as the Akash Ganga Trust, Exnora, Rotary Clubs, Lions Clubs and the Indian National Trust for Art and Cultural Heritage have been playing an active role in promoting RWH in Chennai.

The Tamilnadu government enacted a law in October 2002, and followed with an ordinance in June 2003, making the implementation of RWH systems compulsory in all existing buildings in the entire state of Tamilnadu by 11 October 2003. The law mandates that 'waste water from the bath and wash basin shall be treated by organic or mechanical recycling and taken to a sump for onward pumping to an excessive overhead tank for use in toilet flushing. Any excess shall be connected to the rainwater harvesting structured for groundwater recharge'.

Tamilnadu's *Groundwater Regulation Act 1987* has also been amended to include the power to grant or refuse groundwater licences. Use of groundwater for gardening, private swimming pools and for non-potable uses by industries has been banned. Those who are found guilty are fined Rs 2000 for the first offence and Rs 5000 for the second offence.

A survey was done in the Gandhi Nagar, an area which comprised 309 contiguous plots occupied either by independent houses or multistorey apartment complexes. The survey revealed that all of the 309 residences had installed RWH structures. Another survey examined how citizens managed their water needs for both potable and non-potable use. The findings revealed that people are not only learning to live with less water, but also with different qualities of water for drinking, cooking, bathing, washing and toilet-flushing.

9. Legislation encouraging rainwater harvesting in India

India is the first country to have amended its National Constitution to address the threat of degradation of the environment and natural resources. The 42nd Amendment to the Indian Constitution in 1976 underscored the importance of green thinking. Article 48A enjoins the state 'to protect and improve the environment and safeguard the forests and wildlife in the country'. Further, Article 51A(g) states that the 'fundamental duty of every citizen is to protect and improve the natural environment including forests, lakes, rivers, and wildlife and to have compassion for living creatures'. In the recent past, these constitutional provisions have influenced India's environmental policies. The Government of India has articulated three policy statements, namely the *Policy Statement on Abatement of Pollution* (1992), the *National Conservation Strategy and Policy on Environment and Development* (1992) and the *National Water Policy* (2002).

In the *National Water Policy*, under the subheading 'Water Resources Planning', paragraph 3.2 clearly highlights the importance of rainwater harvesting by stating:

> *Non-conventional methods for utilization of water such as through inter-basin transfers, artificial recharge of ground water and desalination of brackish or sea water as well as traditional water conservation practices like rain water harvesting, including roof-top rainwater harvesting, need to be practiced to further increase the utilizable water resources. Promotion of frontier research and development, in a focused manner, for these techniques is necessary.*

Further, it has been widely accepted by the administrative and regulatory authorities that there should be a periodical reassessment of the groundwater potential on a scientific basis, taking into consideration the quality of the water available and economic viability of its extraction. Exploitation of groundwater resources should be so regulated as not to exceed the recharging possibilities, and also to ensure social equity:

> *In due course, several Memorandums of Understanding were signed between the central government and the various state governments of India which signifies the importance of rainwater harvesting. Paragraph 5 (viii) of one such memorandum states that each water supply scheme will incorporate conservation measures, rainwater harvesting and ground water recharge systems for source sustainability (Department of Drinking Water Supply 2003).*

The report of the Planning Commission's Steering Committee on Drinking Water Supply & Sanitation (Rural & Urban) for the Tenth Five Year Plan signifies the importance and the implementation of rainwater harvesting in the states:

> *Integrated water supply and sanitation programmes will be increasingly implemented during the Plan. The implementing machinery in the Centre and the State will require organisational restructuring to work in a Mission mode with guidance from the Rajiv Gandhi*

National Drinking Water Mission Authority and its empowered committees. Micro water-shed based master plans should be prepared to ensure the sustainability of water sources by taking care of demand and supply. The inputs of professional institutions, NGOs and Community Based Organisations (CBOs) should be inducted into planning, development and management. At the same time, integrated water use and conservation methods should be adopted. (Planning Commission 2002, p. 6)

Other recommendations of the committee include the following:

- *All possible measures for rainwater harvesting and ground water recharging must be taken. There should be continuous monitoring of the sources, so that the habitations once covered do not fall back in the category of uncovered, for which interdepartmental coordination at Block level need be activated.*
- *Water supply links with water-shed development programme [sic] should be [strengthened] for better sustainability of drinking water sources.*
- *Traditional sources shall be identified, strengthened and developed with community involvement. Rehabilitating the existing village tanks, creating detention basins by storing rain water in local depressions, abandoned mines/quarries etc. for water harvesting for the development of water resources [needs to] be encouraged. Small dams should be encouraged, because micro water-shed area is more efficient for water conservation. To avoid the evaporation losses from such small storages, the underground syphon should be used, which would conserve the water and recharge the aquifer. (Planning Commission 2002, pp. 6–7)*

As a result of the various legislation and planning directives, various steps have been taken for the development of the rainwater harvesting in India's states, but a concentrated effort is still required by all states to tackle the problem of water. Some of the developments are listed below (CSE 2005).

9.1 State of Kerala

- The *Kerala Municipality Building Rules 1999* were amended in 2004 to include RWH structures in new construction.
- The new chapter on Rainwater Harvesting in the Rules requires rainwater harvesting arrangements to be provided as an integral part of all new building constructions, whether residential (with a floor space of more than 100 square metres), educational, medical, assembly, business or industrial buildings.

9.2 State of Delhi

- Since June 2001, the Ministry of Urban Affairs and Poverty Alleviation has made rainwater harvesting mandatory in all new buildings with a roof area of more than 100 square metres and in all plots with an area of more than 1000 square metres.

9.3 State of Madhya Pradesh

- Rainwater harvesting has been made mandatory in all new buildings with an area of 250 square metres or more.
- A rebate of 6 per cent on property tax has been offered as an incentive for implementing RWH systems.

9.4 State of Uttar Pradesh

- Rainwater harvesting has been made mandatory in all new buildings with an area of 1000 square metres or more.

9.5 State of Andhra Pradesh

- Rainwater harvesting has been made mandatory in all new buildings with an area of 300 square metres or more.
- The *Andhra Pradesh Water, Land and Trees Act 2002* came into force on 1 July 2002 with the objective of promoting waste conservation and tree cover, and regulating the exploitation and use of ground and surface water. The aim was to protect and conserve water sources, land and environment.

9.6 State of Tamilnadu

- Through Tamilnadu Municipal Laws, Ordinance No. 4 of 19 July 2003, the government of Tamilnadu has made RWH mandatory for all buildings, both public and private, in the state. The deadline given was 31 August 2003.

9.7 State of Haryana

- Haryana Urban Development Authority (HUDA) has made RWH mandatory in all new buildings, irrespective of roof area.
- The Central Ground Water Authority (CGWA) has also banned the drilling of tubewells in notified areas.

9.8 State of Rajasthan

- The state government has made RWH mandatory for all public buildings and establishments, and all properties in plots covering more than 500 square metres in urban areas.

9.9 State of Maharashtra

- The state government has made rainwater harvesting mandatory for all buildings that are being constructed on plots more than 1000 square metres in size.
- The deadline set for this was October 2002.

9.10 State of Gujarat

- The State Roads and Buildings Department has made RWH mandatory for all government buildings.

10. Application of remote-sensing data and GIS in rainwater harvesting

In today's world of information technology, remote sensing and geographical information systems have emerged as particularly useful tools for decision-makers in every field of society. Remote sensing is the science (and to some extent, the art) of acquiring information about the planet's surface without actually being in contact with it. This is done by sensing and recording reflected or emitted energy, and processing, analysing and applying that information. GIS (Geographic Information System) is a computer-based tool for analysing and mapping things that exist and events that happen on the earth. GIS technology integrates common database operations such as query and statistical analysis with the unique visualisation and geographic analysis benefits offered by maps.

While remote-sensing data and GIS are useful tools for implementing a RWH scheme in a particular area, it is important to note that analysis and interpretation of satellite imagery can also help find suitable areas for water harvesting. This, however, should be done carefully and, if possible, in coordination with local people. Sometimes visual interpretation by local land users can be more fruitful than computer-based analyses of spectral data by an external researcher. The involvement of local land users could be termed 'Participatory Image Interpretation'. This holds great promise for dry areas, because the normally low economic value of these areas typically results in a deficit of data about them. Furthermore, because of the large areas of these environments and the relatively low population density, it is very expensive to inventory them using traditional surveying methods.

Recently, remote-sensing data and GIS have been used worldwide for the assessment of RWH potential and requirements for respective regions. In India, too, these decision-making tools are now being used vigorously. For example, in the Kovilpatti area of the Salem district of Tamilnadu, GIS was used to delineate favourable zones for recharge by integrating various thematic map techniques, and recommending suitable recharge structures. This was done by making layers of various maps such as geomorphology, geology, soil maps, slope maps, land use maps and drainage maps. Using these layers, a map showing the best places for locating recharge pits was made.

11. Skill-building workshop on rainwater harvesting

In December 2005, the Indian Institute of Technology, Bombay (IIT) and the Indian Environmental Association organised a two-day skill-building workshop in

Mumbai entitled 'Treatment of Harvested Rainwater for Industrial and Domestic Use'. The workshop aimed to disseminate the outcomes of action-learning projects on RWH executed by the graduate students of the IIT. It also investigated the potential of practising RWH in industry and the community. The participants in the workshop were professionals and practitioners including environmental engineers and scientists, planners, architects, regulators, MCGM officers, students, members of non-government organisations, and concerned citizens who were either already practising or wanted to practise RWH. The workshop had renowned speakers from India and abroad who not only taught basic engineering and science but shared their experiences as well.

The main attraction of the workshop was the two keynote addresses delivered by two eminent personalities. Professor Soli J Arceivala, a practising environmental consultant from India, underscored the need for RWH in India. He strongly recommended RWH projects for an individual household or a group of households in rocky terrain with no groundwater supplementation, and to raise the general aquifer level in an area for improving groundwater availability for a community as a whole. In his concluding remarks, he said, 'If water is running, make it walk; if it is walking, make it stand; if it is standing, make it sit; and if it is sitting, make it sleep!'

The second keynote address was delivered by Professor Margaret Robertson, a noted educationist and geographer from the Faculty of Education, University of Tasmania, Australia. She elaborated on the pedagogical focus on sustainable futures and recommended the collaboration of non-government organisations and schools, together with industry and engineers. Her critical review enlightened the RWH professionals.

At the end of the workshop, Dr Deepak Kantawala, a noted environmental consultant and policy researcher in India, chaired a group discussion and experience-sharing session. It was concluded that RWH projects would become viable in Mumbai only if suitable treatment could be given to the harvested water. This would augment potable and industrial water supplies. In the case of commercial complexes, restaurants and hotels, harvested rain from rooftops and paved surroundings can indeed supplement a part of the water demand because water rates for commercial and semi-commercial applications in Mumbai are high enough to justify investments in RWH systems.

12. Conclusions

The current status of the water crisis, and the situation ahead, have not been brought about by natural factors like drought. It is due to ever-increasing population pressure, urbanisation and industrialisation. According to a 2003 World Bank study of the 27 Asian cities with populations of over one million, Chennai and Delhi are the worst-performing metropolitan cities in terms of hours of water availability per day, while Mumbai is the second-worst performer and Kolkata fourth-worst (Background Paper 2001).

Rainwater harvesting, however, can be a viable solution to the alarming water crisis in both rural and urban areas. It is not only a means to resolve water crises, it is also a measure to eradicate poverty, to generate massive rural employment and to reduce migration from rural and urban areas. It is not only the responsibility of the government, but also the duty of civil society and ordinary citizens to work together to eradicate the problem of water shortage. From the case studies of rural and urban areas it has been proved that community participation is crucial for the success of any project. The role of government is also significant. Government can achieve success in any project through proper management and strict implementation of laws and policies. Rainwater harvesting systems are successful through a joint effort of government, civil society and people.

Appendix 7.1 How to estimate rainwater harvesting potential

RWH potential in India

India is blessed with 113 rivers flowing up to approximately 45 000 kilometres in length, consisting of 14 major, 44 medium and 55 minor rivers. Three of the major rivers are international and the remaining 11 are national, and together they contribute to approximately 80 per cent of the India's water supply.

The average rainfall of India, distributed over the entire Indian geographical area, comes to about 1200 millimetres (against 860 millimetres for the entire world). As India's total geographical area is 328 million hectares (i.e. 3.28 million square kilometres), the total rainfall will be 400 million ha-m (hectare-metre), including water associated with 10 million ha-m of snowfall. The distribution of precipitation is estimated to be as shown in Table 7.1 (page 118).

In addition to this precipitation of 400 million ha-m, 10 million ha-m surface run-off comes from outside India, thus producing a total run-off or stream flow in India of 115 + 45 + 10 = 170 million ha-m. Out of these, only 31 million ha-m are utilised through pumping for irrigation and power as well as storage in dams, lakes and so on, while the remaining 139 million ha-m go to sea. With the proper planning and development of infrastructure facilities, a further 40 to 50 million ha-m of rainwater can be harvested and utilised (Patil 2003).

Estimation of rainwater harvesting potential

The collection and storage of rainwater from run-off areas such as roofs and other surfaces has been carried out since time immemorial. By careful design it is possible

Table 7.1   Distribution of precipitation in India

Distribution of precipitation	Volume (million ha-m)*
Surface run-off	115
Evaporation	70

Percolation:
    base flow for rivers
    groundwater table
    soil moisture
	215
	45
	50
	120

Total	**400**

*   1 ha-m (i.e. 1 hectare × 1 metre) = 100 000 cubic metres (i.e. 100 million litres)

for a family to be sustained for a year in areas with an annual rainfall of as little as 100 millimetres. Many observations made around the world, especially in Zimbabwe, Botswana and Israel, show that between 80 to 85 per cent of all measurable rain can be collected and stored from outside catchment areas. This includes light drizzle and dew condensation, which can occur in many parts of the country during the drier months. The run-off from a catchment area can be worked out by the simple formula:

$Q = C \times I \times A$
Where: Q = discharge in cubic metres
           C = co-efficient of run-off
           I = total rainfall per annum, in metres
           A = catchment area in square metres

The co-efficient of run-off depends upon the shape, size, soil conditions, temperature and geological conditions of the area of the catchment. However, based on average rainfall in the area, the co-efficient can be assumed as follows in Table 7.2.

One millimetre of rainfall collected in one square metre of area will typically yield approximately one litre of water. The minimum requirement of potable water for the purposes of human consumption in extreme drought conditions is typically about 20 litres per capita per day (LPCD) and an additional 30 litres per domestic animal per day. So a family of five, with two bulls and a cow, requires a minimum of 190 litres daily. If we assume that the longest dry period (without availability of water from any source) would be six months, the volume of water required to last through the dry season would be 180 × 190 = 34 200 litres. Assuming an average drought-prone area rainfall of 500 millimetres per annum, 80 per cent capture efficiency of rainwater harvesting from rooftops and 70 per cent storage efficiency in underground storage tanks (after losing water through seepage, percolation and

Table 7.2   Co-efficient of run-off

Condition of catchment area	Annual average rainfall (mm)	Co-efficient of runoff (%)
Dry tracts	350–750	15–20
Intermediate zones	750–1500	20–30
Higher zones	1500	30–55
Roof and paved area		80–90

evaporation), then an area of about 122 square metres of rooftop will be required to harvest 34 200 litres of water.

However, a typical rural household has about 40 square metres of rooftop on a house, and 20 square metres of roof on an animal shelter. Thus, about three months of this family's water requirement (out of six months of the dry season) can be satisfied through rooftop rainwater harvesting. It should be noted that the benefits of RWH can be enjoyed only by those who can spend money on constructing a storage tank of approximately 25 000 litres capacity. Further, communities living in moderately wet areas are habituated to using about three times more water than the people from drought-prone areas. Rooftop rainwater harvesting becomes even less attractive for such communities because even one month's water requirements cannot be satisfied through RWH for them.

In the case of Mumbai, the rainwater harvesting cell of the MCGM has made concerted efforts to promote and disseminate RWH projects all over the Mumbai region. They have published literature in the local language (Marathi) as well as in English. Hundreds of copies have been distributed among the local self-government agencies, non-government organisations, advance locality management groups, architects, plumbers and housing societies. The following approach and calculation is extracted from their booklet (MCGM 2003):

Area of Mumbai	=	437 square kilometres
Annual average rainfall	=	2000 millimetres
Total rainwater falling over Mumbai	=	437 × 1 000 000 × 2000 litres per year
	=	847 000 million litres per year
	=	2394.5 million litres per day
Present water supply	=	2900 million litres per day

If we take the case that only 70 per cent of the total area of Mumbai is to be subjected to RWH, the harvested rainwater works out to be 589.34 million litres per day. This is a sizable quantity compared to the water supply of Mumbai (MCGM 2003).

Appendix 7.2 How to design rainwater harvesting systems

Requirements for designing RWH systems

The secondary data required for the rational design of RWH are listed below.

1   Survey details
    a   plain table survey of the area showing different structures, open places, roads, public gardens, etc.
    b   contour map of the area
2   Rainfall data
    a   annual average rainfall
    b   peak rainfall in 15 minutes in the particular area during rainfall season
    c   the maximum rainfall in a day in that particular area during rainfall season
3   Hydrogeological investigations
    a   the existing pattern of natural drainage of water into subsoil
    b   porosity and permeability of soil and the rock formation below it
    c   data/report from geographical information systems (GIS)
    d   data/report from groundwater supply department
    e   well inventory.

While designing RWH systems we should be aware that there is always some outflow of water from the area under consideration during the rainy season. Percolation systems or storage should be made to absorb and/or accommodate the maximum amount of rainwater/surface run-off from the catchment area.

Design of a rooftop rainwater harvesting system

Table 7.3 Collecting rainwater for drinking and cooking purposes in a tank:

Parameter	Unit
Average number of household members	5
Water requirement for drinking and cooking	10 LPCD
Water requirement for one year	18 250 litres (or 20 cubic metres)
Average area per one rainwater downtake pipe	40 square metres

If we construct a tank of 20 cubic metres, the water collected in the tank will be sufficient for an entire year.

The size of the tank may be one of the following:

1   3.5 metres in diameter and 2.1 metres in height
2   2.5 × 4 × 2 metres
3   3.2 × 3.2 × 2 metres

Assuming only 50 per cent of rainwater is stored in this tank, the annual rainfall which can be accommodated in it would be:

$$2 \times 18\ 250\ /\ 1000 \times 40 = 0.912 \text{ metres (say 915 metres)}$$

Factors affecting the size and number of tanks to be constructed are:

1   annual average rainfall
2   rooftop area
3   number of downtake rainwater pipes
4   number of people to be served.

The number of tanks needs to be increased to match the water requirements for cooking and drinking.

The first spell of rain is utilised to clean the rooftop to avoid dirt and other contaminants from entering the water storage tanks. This assumes that 50 per cent of the available rainwater is taken up for storage purposes and the remainder is allowed to flow to percolation pits or trenches. Therefore the maximum amount of water available from a rooftop is harvested.

Acknowledgment: The material from Meyer et al., 'Where rivers are born: the scientific imperative for defending small streams and wetlands', published by American Rivers and Sierra Club, is used with permission.

Bibliography

Arceivala, S 2005, 'Rainwater harvesting in hilly areas', presentation to Indian Institute of Technology, Bombay/Indian Environmental Association workshop, *Treatment of Harvested Rainwater for Industrial and Domestic Use*, Mumbai, 16–17 December 2005.

—— & Asolekar, SR 2006, *Wastewater treatment for pollution control*, 3rd edn, Tata McGraw-Hill, New Delhi.

Asolekar, S 2002, 'Greening of industries and communities', in LEAD (ed.), *Rio to Johannesburg: India's experience in sustainable development*, Orient Longman, Hyderabad.

—— & Gopichandran, R 2005, *Preventive environmental management: An Indian perspective*, Foundation Books/Centre for Environmental Education, New Delhi/Ahmedabad.

Background Paper 2001, International Conference on New Perspectives on Water for Urban and Rural India, 18–19 September, New Delhi.

Banergee, S 2005, 'Economics of RWH: Case study from Gansu, China', class project, Centre for Environmental Science and Engineering, Indian Institute of Technology Bombay, Mumbai.

Bara, A 2005, 'Traditional rural and contemporary rainwater harvesting systems', class project, Centre for Environmental Science and Engineering, Indian Institute of Technology Bombay, Mumbai.

Bhatia, R 2005, 'Implementation of RWH scheme's perspective of local governing body', presentation to All India Institute of Local Self Government and Indian Water Works Association seminar, *Rainwater harvesting systems: Planning, design, construction, and maintenance*, Mumbai, 28 February–1 March 2003.

Chaturvedi, M, Dwivedi, J, Dixit, S, Singh, M & Gajbhiye, H 1999, 'Dewas as environmental hot spot for water: Assessment and management scenario', class project, Institute of Environmental Management and Plant Sciences, Vikram University, Ujjain.

Chilekar, S 2005, 'Filtration and disinfection technologies for harvested rainwater', presentation to Indian Institute of Technology, Bombay/Indian Environmental Association workshop, *Treatment of Harvested Rainwater for Industrial and Domestic Use*, Mumbai, 16–17 December 2005.

CSE 2005, *Legislation on rainwater harvesting*, CSE, New Delhi, viewed 28 May 2007, (http://www.rainwaterharvesting.org/Policy/Legislation.htm).

Daiya, D 2005, 'Water usage optimization', presentation to All India Institute of Local Self Government and Indian Water Works Association seminar, *Rainwater harvesting systems: Planning, design, construction, and maintenance*, Mumbai, 28 February–1 March 2003.

Department of Drinking Water Supply (DWS) 2003, *Memorandum of understanding between the State Government of ____ and the Department of Drinking Water Supply*, DWS, Delhi, viewed 28 May 2007, (http://ddws.nic.in/Data/Swajal/Draft-MoU.htm).

Deshmukh, A 2005, 'Recycling, reclamation, and reuse of wastewater', presentation to All India Institute of Local Self Government and Indian Water Works Association seminar, *Rainwater harvesting systems: Planning, design, construction, and maintenance*, Mumbai, 28 February–1 March 2003.

Dixit, M & Patil, S 1996, 'Rainwater harvesting: A discussion paper', in WEDC (ed.), *Reaching the unreached: Challenges for the 21st century* (proceedings of the 22nd WEDC Conference, New Delhi, India), WEDC, Loughborough, UK.

Ganapule, R 2005, 'Importance of geography in rainwater harvesting', presentation to All India Institute of Local Self Government and Indian Water Works Association seminar, *Rainwater harvesting systems: Planning, design, construction, and maintenance*, Mumbai, 28 February–1 March 2003.

Gopinath, K 2005, 'Rainwater harvesting methods with relevance to coastal area', presentation to All India Institute of Local Self Government and Indian Water Works Association seminar, *Rainwater harvesting systems: Planning, design, construction, and maintenance*, Mumbai, 28 February–1 March 2003.

Gupta, G 1994, 'Influence of rain water harvesting and conservation practices on growth and biomass production of *Azadirachta indica* in the Indian desert', *Forest Ecology and Management*, vol. 70, pp. 329–39.

—— & Pann, P 2005, 'On site detention systems for rainwater harvesting', presentation to All India Institute of Local Self Government and Indian Water Works Association seminar, *Rainwater harvesting systems: Planning, design, construction, and maintenance*, Mumbai, 28 February–1 March 2003.

Joshi, A 2005, 'Contribution, maintenance, and costing of rainwater harvesting systems', presentation to All India Institute of Local Self Government and Indian Water Works Association seminar, *Rainwater harvesting systems: Planning, design, construction, and maintenance*, Mumbai, 28 February–1 March 2003.

Kakkar, R 2005, 'Issues and technologies for RWH', class project, Centre for Environmental Science and Engineering, Indian Institute of Technology Bombay, Mumbai.

Kshirsagar, S 2005, 'Need and benefits of rainwater harvesting', presentation to All India Institute of Local Self Government and Indian Water Works Association seminar, *Rainwater harvesting systems: Planning, design, construction, and maintenance*, Mumbai, 28 February–1 March 2003.

Marathe, S 2005a, 'Community and government: Joint initiative', presentation to All India Institute of Local Self Government and Indian Water Works Association seminar, *Rainwater harvesting systems: Planning, design, construction, and maintenance*, Mumbai, 28 February–1 March 2003.

—— 2005b, 'Rainwater harvesting techniques and their applicability', presentation to All India Institute of Local Self Government and Indian Water Works Association seminar, *Rainwater harvesting systems: Planning, design, construction, and maintenance*, Mumbai, 28 February–1 March 2003.

MCGM 2003, 'Water conservation and rainwater harvesting for Brihanmumbai', Municipal Corporation of Greater Mumbai (MCGM), Mumbai.

Meyer, J et al. 2003, 'Where rivers are born: The scientific imperative for defending small streams and wetlands', American Rivers/Sierra Club, Washington D.C./San Francisco, viewed 28 May 2007, (http://www.urbanfauna.org/files/whitepaper_on_headwater_streams.pdf).

Mungekar, N 2005, 'Roof water management: holiday home and training centre', presentation to All India Institute of Local Self Government and Indian Water Works Association seminar, *Rainwater harvesting systems: Planning, design, construction, and maintenance*, Mumbai, 28 February–1 March 2003.

Murghan, N 2005, 'Use of R.S. data and GIS in RWH', class project, Centre for Environmental Science and Engineering, Indian Institute of Technology Bombay, Mumbai.

Paranjpe, U 2005, 'Design of various rainwater harvesting systems', presentation to All India Institute of Local Self Government and Indian Water Works Association seminar, *Rainwater harvesting systems: Planning, design, construction, and maintenance*, Mumbai, 28 February–1 March 2003.

Patankar, S 2005, 'Rooftop rainwater harvesting for Mumbai: An additional resource', presentation to All India Institute of Local Self Government and Indian Water Works Association seminar, *Rainwater harvesting systems: Planning, design, construction, and maintenance*, Mumbai, 28 February–1 March 2003.

Patil, S 2003, 'Rainwater harvesting', presentation to All India Institute of Local Self Government and Indian Water Works Association seminar, *Rainwater harvesting systems: Planning, design, construction, and maintenance*, Mumbai, 28 February–1 March 2003.

Planning Commission 2002, *Report of the Steering Committee on Drinking Water Supply & Sanitation (Rural & Urban) for the Tenth Five Year Plan*, Planning Commission, Delhi, viewed 28 May 2007, (http://planningcommission.nic.in/aboutus/committee/strgrp/stg_water.pdf).

Raghavan, S 2004, 'Rainwater harvesting in urban areas: The Chennai experience', *Aridlands Newsletter*, no. 56, viewed 28 May 2007, (http://ag.arizona.edu/OALS/ALN/aln56/raghavan.html).

Rajabu, K 2005, 'The role of participatory problem analysis in performance improvement and sustainable management of rainwater harvesting (RWH) systems: A case study of Makanya Village, Tanzania', *Physics and Chemistry of the Earth*, vol. 30, pp. 832–9.

Samant, H 2005, 'The role of geology in rainwater harvesting', presentation to All India Institute of Local Self Government and Indian Water Works Association seminar, *Rainwater harvesting systems: Planning, design, construction, and maintenance*, Mumbai, 28 February–1 March 2003.

Sharma, N 2005a, 'Legislations encouraging rainwater harvesting in India', class project, Centre for Environmental Science and Engineering, Indian Institute of Technology Bombay, Mumbai.

—— 2005b, 'Rainwater harvesting: A proactive approach', presentation to All India Institute of Local Self Government and Indian Water Works Association seminar, *Rainwater harvesting systems: Planning, design, construction, and maintenance*, Mumbai, 28 February–1 March 2003.

Undegaonkar, M 2005, 'Constituents of harvested rainwater and overview of treatment options', class project, Centre for Environmental Science and Engineering, Indian Institute of Technology Bombay, Mumbai.

Websites

Rainwater harvesting—India

- Centre for Science and Environment (CSE): Rainwater Harvesting, (http://www.rainwaterharvesting.org)
- Development Process of Ralegan as Model Village, (http://www.manage.gov.in/pune/development_process.htm)
- Government of Kerala: Amendment to the *Kerala Municipality Building Rules 1999*, (http://www.rainwaterharvesting.org/Policy/img/kerala_govt.pdf)
- National Water Policy 2002, (http://www.nih.ernet.in/belgaum/NWP.html)
- Tamilnadu Water Supply and Drainage Board: Rainwater Harvesting, (http://www.aboutrainwaterharvesting.com)

Rainwater harvesting—general

- Envireau, (http://www.envireau.co.uk)
- Global Development Resource Center: Rainwater Harvesting, (http://www.gdrc.org/uem/water/rain water/index.html)
- RainSoft Water and Air Treatment, (http://www.rainsoft.com)
- Sustainable Earth Technologies: Rainwater Harvesting, (http://www.sustainable.com.au/rainwater.html)
- The Texas Manual on Rainwater Harvesting, (http://www.twdb.state.tx.us/publications/reports/RainwaterHarvestingManual_3rdedition.pdf)

Development resources

- International Development Research Centre, (http://archive.idrc.ca)
- M.S. Swaminathan Research Foundation, (http://www.mssrf.org)
- The Regional Institute for Appropriate Small Farming and Animal Husbandry, Brazil, (http://www.irpaa.org.br)
- United Nations Environment Programme: Division of Technology, Industry, and Economics, (http://www.unep.fr/en/)

Other information

- International Water History Association, (http://www.iwha.net)

8 | Water resources, tourism and sustainable development in the Tres Palos lagoon area, Mexico

Álvaro Sánchez-Crispín

This chapter examines the relationship between the use of natural resources, in this case water of a fluvial-lagoon system,[1] and the expansion of the tourism economy, based on empirical evidence. The setting of this is a poor area of the Mexican Pacific coast, not far from the world-famous seaport of Acapulco, some 400 kilometres south of Mexico City. The aim of this research is to reveal how the process of pollution of local water resources, essential for the promotion of recreational activities, can negatively affect the sustained practice of tourism where this type of economy is arising as a local source of both employment and income. The key question of this study is framed to demonstrate how frail the liaison between nature and society is, and how this relationship has to be regulated in order to secure better living conditions for the local population, an issue that is central to the idea of sustainability.

There are five major sections in this study:

- an explanation of the concepts of sustainability, sustainable development and sustainable tourism
- a brief account of the tourism economy in recent years, with particular reference to Latin America and Mexico
- a description of the Acapulco region, both in terms of physical and socioeconomic attributes
- characteristics of the study area and the major findings of a questionnaire/survey
- the general conclusions of the project.

1. Sustainability, sustainable development, sustainable tourism

While the concept of sustainability comes originally from the natural sciences and is related to the permanent use of natural resources (Lanza 2002), nowadays it is used as an umbrella term to refer to all human activity. Several 'sustainable' concepts also refer to sustainable development, sustainable growth, sustainable agriculture and sustainable tourism, among many others.[2] Sustainability relates to the continuity of every aspect of human society—environmental, social, economic and institutional—and for this reason, it affects every feature of social organisation, from the local level to the entire planet. In simple words, sustainability aims at providing continuously the best for the environment and people (Young & Hamshire 2000). The concept of sustainability is therefore much more than environmental protection. It is a positive idea concerned with reducing impacts, either environmental or social, for the permanent wellbeing of people and the natural environment.

The very root of the word sustainability, from the verb 'to sustain', meaning to support or to keep going, is basic to understanding why sustainability is about the future. Whatever the academic relevance of sustainability may be, this concept is deeply rooted in human expectancy and ways of life. These two issues heavily influence the whole idea of sustainability at a given place and connect it to what people want the future to be in a given setting, whether that setting is urban, rural, agricultural or industrial. Sustainability is also related to how people have been making a living in particular scenarios for hundreds of years and have established a congruent relationship with nature, which in turn has permitted citizens to use natural resources on a permanent basis. Regardless of the type of environment and the human occupation of it, the analysis of the balance between natural resources exploitation and human demands is central to the idea of sustainability, and it can be studied from different scientific standpoints. Since geography is the study of the interaction between nature and society from a territorial perspective, the role of geographical studies becomes fundamental when interpreting the meaning of sustainability.

While academics and the general public currently use the word 'sustainability', the term 'sustainable development' has a longer history, and dates back to the end of the 1970s. Several definitions of sustainable development arose during the 1980s and 1990s. The most quoted of them is contained in the report of the Brundtland Commission (also known as *Our common future*) first published in 1987 and presented to the United Nations World Commission on Environment and Development. The definition was later modified during the Rio (1992) and Johannesburg (2002) Summits (Beruchashvili, Chauke & Sánchez-Crispín 2004; Houtsonen 2004). In the Brundtland Report sustainable development is defined as development that 'meets the needs of the present without compromising the ability of future generations to meet their own needs' (World Commission on Environment and Development 1987, p. 24).

At present, the concept of sustainable development is still used by many researchers and teachers working in such different areas as ecology, biology, sociology, anthropology and geography. In the case of geography, sustainable development is a paradigm widely employed in a variety of studies, both in physical and human geography. Sustainable development and sustainability itself are firmly anchored in both dimensions of the geographical analysis, with one examining the natural features of places and the other concerned with the study of the human presence on earth.

The main objection to the use of the term 'sustainable development' as an umbrella term is that it implies continued development. Critics of its use insist that it should be reserved only for development activities, not the whole spectrum of human activity. Nonetheless, one popular definition describes sustainable development as 'the improvement of the quality of human life whilst living within the carrying capacity of the ecosystems'. Thus, sustainable development should be a process based on a harmonious use of natural resources in order to meet the demands of the present and the future population at a given place, be it a small village or the whole planet. Ideally, this process should guarantee the constant use of natural resources to satisfy the demands of human society at all times. However, in reality, we are not even meeting our present needs, a situation that brings under scrutiny the very concept of sustainable development.

Stemming from the term sustainable development, several other concepts such as sustainable agriculture, sustainable mining and sustainable tourism have managed to appear in the scientific literature in the last decade. In this context—and central to the development of this study—is the idea of sustainable tourism, defined as an industry that attempts to make a low impact on the natural environment and local culture of the place where tourists are visiting and, at the same time, contribute to the creation of jobs, income and the preservation of nature. It is responsible tourism that is both ecologically and culturally sensitive (Butler 1998). While they are not the same, people tend to use the terms 'sustainable tourism' and 'ecotourism'[3] interchangeably (Honey 1999).

A central premise of sustainability is 'first, do no harm', so therefore it is expected that sustainable tourism should not abuse the destination and, consequently, cannot be associated with the depletion of local natural or cultural resources. When the principles of sustainable tourism are applied, there should be:

- a minimisation of pollution and a rationalisation of energy consumption and water usage
- a respect for local culture and traditions
- the involvement of the community in the tourism economy
- an expectation of quality rather than quantity in terms of the number of visitors arriving at tourist destinations—small-scale tourism is preferred over massive tourism
- economic benefits for the local population rather than for foreign tourism companies.

This probable outcome of implementing sustainable tourism standards is extremely significant for this research since these principles serve as a guideline for two things: firstly, the evaluation of the relationship between nature and society in the study area and, secondly, the role such relationships play in the actual shaping of tourism.

Added recently to the sustainability jargon is the concept of geotourism, proposed by National Geographic.[4] It refers to the principles of sustainable tourism by building on geographical character or sense of place. Geotourism emphasises the distinctiveness of a locale which is beneficial to visitor and resident alike. This brings about the idea of a type of tourism focused on the uniqueness of a place, seen as the characteristics that make it distinctive such as the physical environment, human presence (culture, heritage), aesthetics and the wellbeing of the local population. Thus, geotourism involves the community, informs the tourists about the place they are visiting, brings about economic benefits, supports the integrity of the place and leads to rewarding tourist experiences. Geotourism presents itself as a new trend in travel, shares a wide variety of features with sustainable tourism and is a vision for the future of tourism. It is therefore clear that geotourism has many geographical components.

2. The tourism economy

In the last two decades, the world's tourism economy has been expanding steadily. Some of the reasons include:

1   **The availability of free time and money,** mainly in rich countries, that motivates millions of people to travel to places far and near, in order to consume and enjoy different tourist products such as a tropical beach, culinary delicacies, archaeological sites, a volcano, peculiar flora and fauna or a natural protected area.

2   **The advantages of modern transportation** that allow tourists to move more rapidly and securely to and within different settings: arid, arctic, tropical, coastal, mountainous, rural and urban. One consequence of this is that more and more countries have become tourist destinations, which in turn makes competition in the international tourist market more fierce. Countries that offer singular and uncommon tourist products such as remote islands, glaciers, moon-like landscapes, spectacular waterfalls, endemic fauna and so on are leaders in this area of tourism.

3   **Tourism has become a major employer** in places where the national economy leaves only a few options open for the local population. Thus it is easier to look for a job in a bar, restaurant, hotel or any other service catering to international and national visitors than being employed in agriculture, the manufacturing sector or mining. This is particularly true of poor countries with severe problems such as migration and high unemployment, such as in Mexico.

Today, tourism is a major employer in several nations of the world, both rich and poor. Poor countries with large tourism economies constitute a suitable scenario for the analysis of the relationship between the natural environment and tourism, since most of the tourist products offered in these places are linked to the existence of particular and unique natural resources such as volcanoes, lagoons, rapids, beaches, wild fauna and endangered flora. This type of tourist offer is often called 'ecotourism',[5] a name that automatically and romantically evokes uncompromising preservation of all natural resources being promoted to lure tourists to a particular area. It is in this context that several Latin American governments are now making every effort to attract the largest number of tourists, and therefore the largest amount of foreign exchange, possible. Such revenue income is needed to boost national earnings, create jobs, and promote regional as well as national development.

The most sophisticated tourists are willing to travel to exotic destinations as long as the tourist offer is unique and remarkable. They do not want to take part in massive tourism ventures or end up in the most hideous tourist traps. Thus, new trends in international tourism favour unspoiled places, in environs that are both beautiful and fragile at the same time. Latin American countries, like Costa Rica, Panama, Venezuela and Ecuador, have taken advantage of the singularity of their own natural resources in order to secure a niche in the world tourism market by declaring large portions of their territory as natural protected areas. In these areas, new forms of tourism, such as birdwatching and the observation of the rainforest from above, can be practised. One benefit to the country is that this type of tourism conserves the natural vegetation and protects valuable water resources. However, in spite of all the apparent advantages of so-called ecotourism, in most Latin American countries the endorsement of old forms of tourism, mainly mass and beach-oriented tourism, still holds sway.

While some new forms of promoting tourism have been introduced to different parts of Mexico (particularly in forested areas and in some river basins), old-fashioned tourism still reigns in the country. With over 20 million international visitors annually, Mexico is among the top ten international tourist destinations in the world and the most important one in Latin America, with its tourists representing half the total number of tourists who arrived in the region throughout the 1990s. Around 80 per cent of the international visitors to the country come from the United States, many of whom are of Mexican American origin (WTO 2004). Once they reach the country, most of these visitors are taken, in large numbers, to the legendary beaches of the Caribbean and Pacific seafronts. This form of tourism has been present in Mexico since the early 1940s, a time when World War II was at its height and the US elites needed places to escape. This is how places like Acapulco appeared on the international tourist map.

These old forms of tourism are still the basis for the promotion of this type of economy in different coastal areas of Mexico. The endorsement of tourism has brought about a constant flow of foreign exchange that, arguably, has benefited the economy and society of particular regions of southern and eastern Mexico.[6] However, over the years the burden imposed on the natural environment by the unregulated construction of hotels, time-shared properties, service areas and infrastructures needed to meet the demands of an ever-growing tourist sector has been enormous, because tourism is one of the least regulated industries in the country. In this context, resources have been exploited without any restrictions. The negative balance resulting from this misuse of natural resources leads to the question of the sustainability of tourism in Mexico. This relationship between nature and tourism can be summed up in many parts of the country as 'harsh', making the issue of sustainability more like part of a fantasy than an achievable, real objective necessary for the development of particular regions. Therefore, our own research in a poor area of Mexico aims to contribute to a wider understanding of the concept of sustainable development, based on the consideration of two factors: water as part of the natural resources endowment in the study area, and the tourism economy expanding from an old tourist centre, Acapulco, that demands the exploitation of those water resources.

3. A geographical overview of Acapulco and its immediate region

Over the last 60 years, the tropical seaport of Acapulco on the Pacific coast of southern Mexico has attracted a large number of tourists, both national and international. While the number of foreign visitors has clearly diminished, from a peak of two million visitors in 1980 to only 600 000 in 2002 (as a consequence of fierce competition with other tourist destinations in the country), Acapulco is still alluring to many tourists, particularly wealthy Mexicans living in Mexico City, a city of more than 20 million inhabitants located 370 kilometres north of this coast. For many years this large number of tourists has placed a growing burden on the environment of the Acapulco region, particularly concerning water resources.

The physical environment of this urban centre includes one of the largest bays in Mexico whose maximum width is six kilometres; a very narrow coastal plain; the impressive mountains of the Sierra Madre del Sur very close to the coastline, reaching more than 2000 metres high a short distance from the coast; and two big lagoons—Coyuca and Tres Palos (Figure 8.1, page 132). The tropical climate of Acapulco, with a marked rainy season similar to the one prevailing in certain parts of South-East Asia, and Central and South America, assures sunshine every day from November to May. These climate conditions have been the basis for the promotion of the local tourism industry. There are two major river basins in the region. The closest to the urban area of Acapulco is the La Sabana River which, with the Tres Palos lagoon, constitutes a fluvial-lagoon system. The other basin is occupied by the

Papagayo River whose main course discharges into the Pacific Ocean very near the Tres Palos lagoon. Abundant tropical vegetation still covers most of the slopes of the Sierra Madre del Sur which face the sea, particularly the northern and eastern flanks. The idyllic mixture of palm trees (most of which are part of the agricultural landscape), golden beaches and perfect weather once generated massive numbers of tourists for many decades. However, this pattern has started to prove unsuccessful now that there are environmentally friendly tourist products in other parts of Mexico and these are now attracting more international visitors than Acapulco.

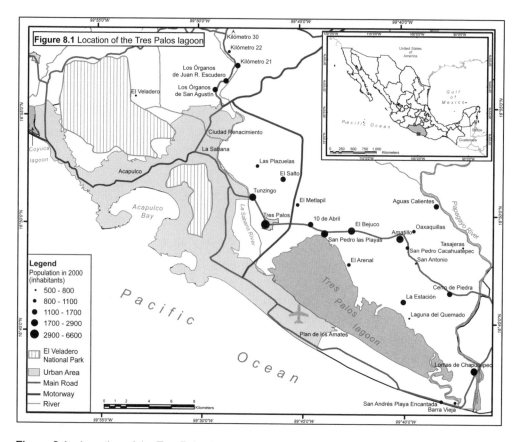

Figure 8.1   Location of the Tres Palos lagoon

Source: Adapted from INEGI (2000)

There is a peak season for international tourism in Acapulco that coincides with winter and early springtime in the northern hemisphere (January to April), including the ever-popular spring-break holidays. In contrast, national tourists arrive during weekends, Easter, summertime holidays (July and August) and at the end of the year. This type of seasonal tourism creates several problems for the local economy since it is only at particular periods of the year that services and other activities associated with tourism are in heavy demand, with negative effects for the local population in terms of employment and income. As the high season for tourism in the Acapulco region coincides with winter and spring, pressure on local water resources increases at a time when natural water availability is at its lowest, due to the climate of the area. In view of these two situations, the potential of sustainable tourism to exist in Acapulco comes under question.

From a shabby town of scarcely 40 000 people in the 1950s, Acapulco has rapidly grown to be one of the largest cities in Mexico. It now has a population of over 600 000 inhabitants, sprawled over an area of 600 square kilometres (see Table 8.1).

The expansion of Acapulco has occurred mainly within the administrative area of its own municipality and in the last decade has absorbed and incorporated many

Table 8.1   La Sabana River–Tres Palos lagoon area: population evolution, 1970–2000

Settlement	1970	1980	1990	2000
Kilómetro 30	3090	5046	4786	5988
Kilómetro 22	n/a	n/a	458	547
Kilómetro 21	594	1183	1772	1347
Los Órganos de Juan Ranulfo Escudero	830	1339	1280	1722
Los Órganos de San Agustín	843	n/a	1273	1336
Acapulco	174 378	301 902	515 374	620 656
Tunzingo	516	576	1324	1828
Tres Palos	2419	2267	4161	4532
10 de Abril	n/a	n/a	615	1168
San Pedro las Playas	425	1089	2025	2906
El Arenal	n/a	336	480	888
La Estación	484	558	918	1304
Laguna del Quemado	205	310	450	525
Lomas de Chapultepec	1221	1265	1725	1977
Plan de los Amates	1148	1169	1296	**
San Andrés Playa Encantada	n/a	45	434	933
Barra Vieja	442	598	916	774
Total	186 595	317 687	539 287	648 431

** Plan de los Amates is now part of Acapulco's urban area.

Source: INEGI (2007)

towns and villages that in the past were physically separated from the urban area. The direction of this process is clearly towards the east of Acapulco where terrain conditions are excellent for real estate development, particularly expensive projects that demand large areas of land and water. One border of the city has already reached the western and southern margins of the Tres Palos lagoon, mainly in the vicinity of Acapulco's international airport (see Figure 8.1). In total, there are 17 settlements in the study area which have a population of over 500 inhabitants each, located either along the west bank of the La Sabana River or near the Tres Palos lagoon's shore (Table 8.1).

The largest settlement in the study area is Acapulco with 620 000 inhabitants, the second largest urban agglomeration in tropical Mexico (Sánchez-Crispín & Propin 2001).[7] This city had a demographic growth rate of over 20 per cent from 1990 to 2000, clearly an extraordinary increase. The other 16 settlements have a combined population of nearly 28 000, of which nearly 38 per cent (over 10 000 inhabitants) live in or near the shores of the lagoon. There are examples among these places where population growth has been rather extraordinary, such as in the San Andrés Playa Encantada area, at the southernmost end of the lagoon, where demographic growth reached almost 115 per cent in the last decade. Other examples of this can be found in the towns of El Arenal and 10 de Abril. While there are affluent neighbourhoods in the urban area of Acapulco, living conditions for the majority of the local population are rather unsatisfactory (INEGI 1998). This state of affairs can be taken as evidence of how tourism has not resulted in an improvement of conditions for most residents in the area.

While tourism is the basis of Acapulco's economy (Propin & Casado 2004), other activities are both significant employers and sources of contamination (see Table 8.2 and Figure 8.2, page 136). For instance, there is a bottling industry whose only plant (a subsidiary of Coca-Cola) is located not far from La Sabana River. Production at this facility is high since soft drinks and bottled water are always in demand by tourists and residents alike. Consequently, the generation of unwanted water with a high content of caustic soda is notable, and the only way to dispose of it is by direct discharge into the river. Cement plants are part of the urban economy of Acapulco and their products are in high demand due to the increasing number of construction sites all over the study area. The disposal site of their waste has been the main course of La Sabana River.

Owners of some tracts of land have resisted pressure from real estate agents and commercial enterprises and have managed to continue with agricultural and livestock activities. There are large properties, in the fringe of Acapulco's urban area, occupied by palm trees for the production of copra and the raising of cattle, particularly on the southern edge of the Tres Palos lagoon. Fertilisers as well as chemicals are used in these properties and the remnants of these are released into the

Table 8.2   La Sabana River-Tres Palos lagoon area: main sources of contamination, by settlement, 2005

Settlement	Major economic activities	Type of waste generated
Kilómetro 30	Agriculture (maize, coconuts) and cattle-rearing	Sewage, solid waste, residual materials from agriculture
Los Órganos de Juan Ranulfo Escudero	Small-scale mining, rudimentary agriculture and limited cattle-ranching	Sewage, solid waste, agrochemicals
Ciudad Renacimiento*	Slaughterhouse facilities, unregulated construction, formal and informal services, commerce, cement industries	Slaughterhouse waste, sewage, rubbish, industrial discarded waters
La Sabana*	Bottling industry, construction, commerce, formal and informal commerce and services	Wastewater from bottling plant, rubbish, sewage
Tunzingo	Soft-drink industries, small-scale production of sweeps, construction, commercial agriculture (coconuts, tamarind, limes) and livestock, commerce, services	Sewage, agrochemicals, rubbish, industrial waste (from the bottling of soft-drink)
Tres Palos	Traditional agriculture, fishing, services	Agricultural and fishing waste, sewage, rubbish
San Pedro las Playas	Fishing, commercial agriculture (maize, coconuts, mangoes), commerce, services	Agricultural, livestock and fishing waste, rubbish, sewage
El Arenal	Fishing, traditional agriculture, commerce	Organic waste, rubbish, sewage
La Estación	Fishing, traditional agriculture, commerce	Organic waste, rubbish, sewage
Lomas de Chapultepec	Fishing, agriculture (coconuts, mangoes, tamarind, limes), commerce	Agricultural and fishing waste, rubbish, sewage
Plan de los Amates*	Fishing, commercial agriculture (coconuts, mangoes), livestock, commerce	Agricultural, livestock and fishing waste, rubbish, sewage
Barra Vieja	Fishing, small-scale commerce, services	Organic waste, rubbish, sewage

* These settlements are now part of Acapulco's urban area.

Source: Arredondo, Ponce et al. (1999) and data-gathering during fieldwork in late 2005

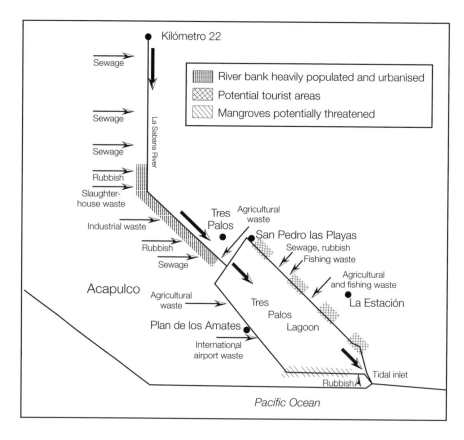

Figure 8.2   La Sabana River–Tres Palos lagoon: main sources of water pollution

Source: Fieldwork data, November 2005

main course of the river or into the lagoon. While there are water treatment plants in the vicinity of the La Sabana River, sewage coming from several settlements located on its western bank and around the lagoon is discharged directly into the lagoon.

In the past, the colour of the lagoon water was bluish-green, but today it has turned brownish and there is a particular smell coming from the water. The lagoon shores were once white and spotless, but nowadays they are spoiled with lots of rubbish, both natural and human-made. However, less polluted waters are found towards the central part of the lagoon where fishing is practised under more natural conditions. Considering the existence of these pollution sources and that the river and the lagoon are part of a single system, which assures the transmission of contaminated waters all along it, it is safe to say that this fluvial-lagoon complex has been used more like a waste-dumping site than a valuable natural resource for the promotion of tourism.

As stated above, tourism is the main source of employment and income for Acapulco and its surrounding area. It has been also mentioned that, in the last few decades, massive tourism has been the prevailing way of promoting this type of economy and its environmental and social consequences are well established. In the last few years, however, and in view of this situation, the private and public sectors, as well as local authorities, have initiated actions to promote other forms of tourism in the area. Despite the fact that visitors interested in this new way of advertising Acapulco are still few, publicity relating to the archaeological sites at La Sabana and Palma Sola, the restoration of the old Spanish fort of San Diego—an outstanding massive building of the sixteenth century—and the exploration of tropical areas north-west and south-east of the city are already underway. Through these experiences it is hoped that tourists can get closer to nature and the local culture. One of the areas promoted for the development of low-impact tourism is the Tres Palos lagoon area.

4. Sustainable development at the Tres Palos lagoon area

Among the most notable features of the physical geography in the Acapulco region is the Tres Palos lagoon, located 25 kilometres east of Acapulco Bay. This body of water has extraordinary natural beauty, and houses a great variety of marine vegetation and wild fauna. It covers nearly 52 square kilometres, being 17 kilometres long and 7 kilometres wide, with a mean depth of 3.8 metres (Arredondo, Ponce et al. 1999). A meandering canal, over 10 kilometres long, connects this lagoon to the Pacific Ocean through a tidal inlet. The movement of waters from the lagoon to the sea is sporadic and erratic. The La Sabana River is the main tributary to this lagoon. The whole system is now in a phase of increasing accumulation of sediments and organic waste, coming from industries located upstream, and wastewater of peripheral settlements in north-eastern Acapulco, which has contributed to the problem of natural drainage and contamination of the lagoon. As a result, the environmental quality of the water has deteriorated and the biodiversity of the lagoon ecosystem is being jeopardised.

This body of water is excellent for viewing wildlife. Tropical birds, fish and other aquatic and terrestrial species dwell and breed in this freshwater ecosystem. Some studies made during the 1980s and 1990s found that there were over 5000 birds nesting in the Tres Palos lagoon area, mainly olivaceous cormorants (*Phalacrocorax olivaceus*), great egrets (*Casmerodius albus*), night herons (*Nycticorax nycticorax*) and white herons (*Egretta thula*) (Arredondo, Quiñónez et al. 1999; Yokoyama-Kano 1981). Natural vegetation in the area surrounding the lagoon includes seasonally dry tropical forests with plants—mainly acacias—reaching 4–5 metres in height. Along the shores of the Tres Palos lagoon there are areas covered with mangrove[8] (particularly white mangrove), carrizo, and other aquatic and halophytic vegetation. Mixed with the natural vegetation of the area are coconut palm, tamarind and mango tree plantations in the vicinity of the lagoon.

Just over 10 000 people live in close proximity to the Tres Palos lagoon. San Pedro las Playas is the largest of all the settlements in the area, with a population of nearly 3000 inhabitants (see Table 8.1). Infrastructure and services at these places are rather limited and most people travel to Acapulco looking for specialised shops, medical and educational services, and long-haul transportation, suggesting that the Tres Palos area is part of Acapulco's commercial and service hinterland. Fishing has been the largest employer in the lagoon area, with some agricultural activity complementing it. A total annual catch of 500 metric tonnes of fish is common. The main species caught are bagre, langoustine, anchovy, sardine and shrimps (Acosta 1984). At present, there are more than 1000 fishermen associated in 22 local and regional fishing cooperatives that provide technical and some financial support to their affiliates.

According to the Urban Plan for the Acapulco Metropolitan Area, the Tres Palos lagoon has good potential for ecotourism projects (Gobierno del estado de Guerrero 1998), particularly at the northern shoreline around the settlements of San Pedro las Playas and La Estación. This governmental consideration has been taken as a good platform for the promotion of tourism in the area. Consequently, from the beginning of this century, the tourism economy has become a major source of employment for the residents around the lagoon. The number of people now engaged in the tertiary sector, especially commerce and services, has increased. The relationship between primary-sector activities and tourism can also be seen in the case of fishing, which aims at providing fresh seafood to markets, hotels and restaurants in the Acapulco region. Several types of fish and crustaceans are in reasonably good demand in this international tourist destination, and local fishermen aim to sell their daily catch in that market. Fishermen also offer their boats to tourists willing to explore the inner part of the lagoon, thus serving the dual purpose of fishing and tourism.

This recent expansion of tourism in the Tres Palos lagoon area has to be examined in light of the principles of both sustainable development and the economic growth expected for the local population which provides services to tourists. Therefore, it was important to assess how local residents perceive, understand and evaluate the lagoon in terms of its environmental problems, employment and income, and as a valuable natural resource that could be used continuously for tourism in the foreseeable future. To achieve this, a questionnaire-based survey was designed and carried out in the study area in October 2005.[9] A total of 100 questionnaires were returned and the main results are presented below.

The largest proportion of those who responded to our questionnaire were born in several towns around the lagoon, and those who are not native have been living in the area for over 25 years. In view of this, it is safe to say that all respondents have a significant knowledge and a well-founded perception of the environs and its socioeconomic problems. Most of the interviewees (87 per cent) were middle-aged men, as many women were reluctant to participate in the survey. Almost half the interviewees have either no formal education or scarcely completed primary school. The same proportion of people has a job in the primary sector, specifically fishing, and they either run their own business or work for small enterprises or cooperatives.

When asked about the quality of the water at the lagoon, half of the respondents pointed out that a 'bad to very bad' quality is a peculiarity of this body of water, while only a minority of the interviewees thought that the quality of the lagoon water was 'good to very good'. Eighty per cent stated that pollution is the main problem of the Tres Palos lagoon. Moreover, 20 per cent of the interviewees thought that tourism was important for the economic development of the area, and nearly 50 per cent considered that the landscape is outstanding, no matter how contaminated the lagoon and its waters are. These comments have to be taken as an indication of how local people appreciate the place in which they live and its natural potential for the development of the tourism economy. The awareness of environmental problems did not deter respondents stating that the Tres Palos lagoon is already a tourist site— as three-quarters of the interviewees did—even if adequate infrastructure is not yet available. Additionally, half of the interviewees stated that fishing is the principal local employer at the Tres Palos lagoon area. In view of this, it is possible to say that both fishing and tourism are at the basis of the economy of the lagoon area, and that they both depend heavily on the quality of water in the lagoon for their sustainability.

Interviewees were also asked about how they perceived the lagoon's attributes. The most frequent answer in this respect was that the lagoon is 'beautiful' (19 per cent), it is 'big' (15 per cent), it is 'filthy' (14 per cent) and it is 'nice' (9 per cent). Several noted that it is a rich place in the sense of biodiversity, the landscape is stunning, it is too abandoned and it is one of the best freshwater lagoons in the country, among other responses. These attributes are not as contradictory as they may seem,[10] because they reflect the perception people have about the potential of the lagoon as a tourist destination and the actual state of affairs regarding the environmental quality of the area. Respondents said that tourists who arrive in the area to appreciate a 'beautiful' and 'big' lagoon are most often Mexicans from Mexico City and Acapulco, with just a few foreigners during the year. Even if some people do camp in the vicinity of the lagoon, most of these tourists stay overnight in Acapulco, as the lagoon area has little to offer in of the way of accommodation and other facilities. These visitors go to the lagoon to explore it and to observe the local fauna and, by doing so, they are the closest version of ecotourists in the study area.

Two issues were included in the questionnaire survey concerning the interviewees' expectations for tourism in the area. One referred to the kind of business that they would open, if they had the chance, in the Tres Palos lagoon area. A large majority of the respondents stated they would rather open and run a restaurant, which reveals the expectations interviewees have about tourism. When asked about the location of a new business, the respondents stated that the two largest towns in the area (San Pedro las Playas and Tres Palos) were preferred, but if they had the chance they would opt for Acapulco. This perception can be considered as an indication of how local people evaluate the towns where they live, and the trust they have in them as suitable places for the development of business linked to the tourism economy.

Finally, two more points relevant to the understanding of sustainable development in the study area are important: first, the measures aimed at resolving pollution, the crucial problem in the area. To clean up the lagoon, four measures were mentioned more frequently by the respondents than any others: carrying out the actual physical cleaning of the lagoon, supported by local residents and the government alike (32 per cent); avoiding sewage problems by controlling the discharge of unwanted water coming from domestic and industrial sources upstream (25 per cent); encouraging community participation and organisation (17 per cent); and establishing preventive measures (15 per cent). These answers account for nearly 90 per cent of all replies.

The cleaning of the lagoon waters seems to be the central issue in trying to achieve sustainable development in the area, since fishing and tourism, the two largest local employers, need better environmental conditions to secure an adequate performance and provide continuous jobs. Thus, it is safe to say that the ideas of preventing, organising and acting are firmly rooted in the interviewees' minds. This shows the willingness of residents to act immediately to solve the environmental problems of the lagoon, but this does not necessarily mean that authorities are also eager to collaborate in this respect. If environmental problems are solved, in one way or another, then a start can be made to reach a more satisfactory ambience in which employment, income and, hopefully, better living conditions for the communities around the lagoon can be guaranteed and maintained in the long run. This would meet one of the central premises of sustainable development.

The second issue is associated with actions that should be taken for the development of the lagoon. According to the questionnaire results, three answers were commonly given: improving the fishing economy at the lagoon (36 per cent), publicising the area and creating a better tourist infrastructure in it (21 per cent), and raising consciousness among locals and visitors with regard to the cleanliness of the lagoon (18 per cent). The rest of the answers, representing a quarter of all respondents, quoted a great variety of ideas, including inviting the local, municipal and state governments to participate in the promotion of the lagoon for tourist purposes, and suggestions about connecting the lagoon to the Papagayo River in order to sanitise the area. Again, local residents revealed their knowledge regarding the importance of fishing and tourism in the study area, and their willingness to implement actions leading to the cleaning of the lagoon activities.

5. Conclusions

In the past decade, the growth of the tourism economy in Acapulco has resulted in an uncontrolled urban expansion over sensitive areas such as the Tres Palos lagoon. In this context, much pressure has been put on local natural resources, particularly land and water. Recreational use of these resources, if not strictly regulated in the following years, will have a devastating effect on marine and terrestrial species living in the lagoon environs. Animal life needs to be protected to allow a sustained development of fishing and tourism, the main economic activities of the communities near the lagoon.

One major environmental problem in the study area is water degradation in the La Sabana River–Tres Palos lagoon system. Different sources contribute to this process, including:

- the use of fertilisers and pesticides on agricultural lands located up the river basin and in the vicinity of the lagoon
- the solid waste of industries and a slaughterhouse found in the middle section of the La Sabana River
- the expansion of Acapulco's urban area demanding more and more space for the construction of residential areas and facilities including shopping malls, hospitals and new roads
- the growth of the tourism economy, associated with an increasing demand for land for the construction of hotels, time-shared properties, mansions and condos. Services and commerce associated with tourism also require space. All these facilities have occupied land, transformed the original landscape and altered the natural dynamics of the lagoon. Also, the demand for drinking water, and water for recreational purposes, has been steadily increasing.

The contamination of the lagoon waters prevents local people and tourists enjoying bathing and other leisure activities. Health problems such as infections of the ears, nose, and throat, as well as stomach and intestinal diseases, can result from swimming in polluted water. Wastewater treatment plants are either insufficient or non-existent along the course of the La Sabana River, and this has a negative impact downstream on the quality of water in Tres Palos lagoon and the ecosystems associated with it. This process is aggravated if one considers that several local communities along the lagoon do not have any sewage system at all and release wastewater directly into the lagoon. There is also the problem of rubbish being dumped into the La Sabana River which adds heavily to the contamination of the whole system. Uncontrolled or inadequate disposal of solid waste is a serious problem in the area and a major contributor to the current unhealthy situation in the Tres Palos lagoon, which hampers the expected development of the tourism economy in the foreseeable future.

One of the typical determinants of sustainability and sustainable development is the control of pollution. To reach this point in the La Sabana River–Tres Palos lagoon system, several actions could be taken.

- Physically clean the area, to get rid of both rubbish and contaminated water. This will then improve the appearance of the lagoon, in terms of the colour and smell of the water and the look of its shores, and will bring about better living conditions for humans and animals.

- Inhibit polluted water flowing into the La Sabana River, and therefore the Tres Palos lagoon, by installing more treatment plants along the upper and middle sections of the river.
- Launch an educational program, integrating public education and the interests of the local tourist industry, to engage in the new forms of promoting the place and attract visitors who are conscious of the environmental and socioeconomic problems faced by local residents.

Hopefully all local residents should be aware of the benefits that the cleaning of the lagoon and the prevention of water pollution may bring about. In particular, young scholars should be enthusiastic about the promotion of local projects intended to improve the natural and socioeconomic conditions in the area. Community participation and organisation is central to the successful completion of those projects. Since tourists tend to rate destinations on the general quality and extent of their physical facilities and the natural and cultural attributes, the Tres Palos lagoon area, if it is to become a primary tourist destination, has to undergo a phase of rehabilitation and promotion of its geographical attributes, understanding these as the sum of the natural and cultural resources it possesses, in order to succeed in a highly competitive market. Residents and visitors alike should take responsibility for the sustainability of the area, a process that is deeply rooted in its quest to endorse responsible fishing and tourism in the long term.

Acknowledgments

I want to thank the following institutions and persons whose help was fundamental to the completion of this research project:

- the International Geographical Union (IGU), its President Professor Adalberto Vallega and its Secretary-General Professor Ron Abler. Their support (financial and professional) was generously given and made possible my attendance at the Rome workshop in June 2005
- Professor Margaret Robertson, whose enthusiasm for the whole project and the coordination of the Rome workshop were transmissible to my own research
- Doctor Nau Niño, a geographer based in Acapulco, whose knowledge of the social network was essential for the development of the questionnaire survey, and the in-depth review of particular socioeconomic processes
- Gerardo Mollinedo and Miguel Angel Garrido, two of my geography undergraduates who helped me with fieldwork
- The Instituto de Geografía at the National University of Mexico, my working base, which provided complementary financial support.

Bibliography

Acosta, R 1984, *Diagnosis de la pesca comercial de la Laguna de Tres Palos, Guerrero, México* [*Diagnosis of commercial fishing at Tres Palos lagoon, Guerrero, Mexico*], Tesis de licenciatura en Biología, Facultad de Ciencias, UNAM, Mexico City.

Arredondo, J, Ponce, J, Luna, C, Coronel, C & Palacios, C 1999, 'Manejo extensivo e intensivo de la pesca en la laguna de Tres Palos' ['Extensive and intensive fishing management in Tres Palos lagoon'], in A Banderas (ed.), *Diseño de las medidas de biorremediación y saneamiento de la laguna de Tres Palos, Acapulco, Guerrero*, vol. II, Instituto Mexicano de Tecnología del Agua, Mexico City, pp. 2–70.

Arredondo, J, Quiñónez, R, Luna, C, Coronel, C & Palacios, C 1999, 'Cinturón de conservación ecológica, tipos de vegetación y fauna en la laguna de Tres Palos' ['Ecological conservation buffer zone and vegetation and fauna types in Tres Palos lagoon'], in A Banderas (ed.), *Diseño de las medidas de biorremediación y saneamiento de la laguna de Tres Palos, Acapulco, Guerrero*, vol. II, Instituto Mexicano de Tecnología del Agua, Mexico City, pp. 287–315.

Banderas, A 1999, 'Caracterización y diagnóstico general del ecosistema río La Sabana-laguna de Tres Palos' ['Characterisation and diagnosis of the La Sabana River-Tres Palos lagoon ecosystem'], in A Banderas (ed.), *Diseño de las medidas de biorremediación y saneamiento de la laguna de Tres Palos, Acapulco, Guerrero*, vol. I, Instituto Mexicano de Tecnología del Agua, Mexico City, Mexico, pp. 2–32.

—— (ed.) 1999, *Diseño de las medidas de biorremediación y saneamiento de la laguna de Tres Palos, Acapulco, Guerrero*, vols. I–III, Instituto Mexicano de Tecnología del Agua, Mexico City.

Butler, R 1998, 'Sustainable tourism—looking backwards in order to progress?', in Hall, CM & Lew, AA (eds), *Sustainable tourism: A geographical perspective*, Longman, Kuala Lumpur, pp. 25–34.

García, E 2004, *Modificaciones al sistema de clasificación climática de Köppen* [*Modifications to Köppen's climate classification system*], Instituto de Geografía, UNAM, Mexico City.

Gobierno del estado de Guerrero 1998, *Plan director de la Zona Metropolitana de Acapulco* [*Urban plan of the Acapulco Metropolitan Area*], Memoria tecnica, Acapulco.

Hall, CM & Lew, AA 1998, *Sustainable tourism: A geographical perspective*, Longman. Kuala Lumpur.

Honey, M 1999, *Ecotourism and sustainable development: Who owns paradise?*, Island Press, Washington DC.

Houtsonen, L 2004, 'Education for sustainability', in B Miranda, F Alexandre & M Ferreira (eds), *Sustainable development and intercultural sensitivity: New approaches for a better world*, Universidade Aberta, Lisboa, pp. 19–32.

INEGI 1998, *Estructura económica del estado de Guerrero* [*The economic structure of the state of Guerrero*], Instituto Nacional de Estadística, Geografía e Informática, Aguascalientes.

—— 2000, *Acapulco de Juárez. Cuaderno Estadístico Municipal* [*Statistical book of the municipality of Acapulco de Juarez*], Instituto Nacional de Estadística, Geografía e Informática, Aguascalientes.

—— 2007, *Instituto Nacional de Estadística Geografía e Informática*, INEGI, Aguascalientes, viewed 29 June 2007, (http://www.inegi.gob.mx/inegi/default.aspx).

Lanza, A 2002, *Lo sviluppo sostenibile* [*Sustainable development*], Societa Editrice Il Mulino, Bologna.

Miranda, B, Alexandre, F & Ferreira M (eds), *Sustainable development and intercultural sensitivity: New approaches for a better world*, Universidade Aberta, Lisboa.

Propin, E & Casado, J 2004, 'Caracterización y diagnóstico económico del estado de Guerrero' ['Economic characteristics and diagnosis of Guerrero state'], in *Caracterización y diagnóstico de la mesorregión sur-sureste y de los estados de Yucatán y Guerrero para el Ordenamiento Territorial*, Instituto de Geografía-UNAM y Secretaría de Desarrollo Social, Mexico City.

Purvis, M & Grainger, A (eds) 2004, *Exploring sustainable development: Geographical perspectives*, Earthscan, London.

Sánchez-Crispín, A & Propin, E 2001, 'Cambios en la orientación funcional de las ciudades medias del trópico mexicano' ('Functional-orientation changes in the medium-sized cities of tropical Mexico'), *Cuadernos Geográficos*, 31, Universidad de Granada, Granada, pp. 69–86.

Vargas, S, López, E & Romero, R 1999, 'Contexto social para el saneamiento de la cuenca del río La Sabana-laguna de Tres Palos' ['Social context for the development of sanitary actions in the La Sabana River-Tres Palos lagoon basin'], in A Banderas (ed.), *Diseño de las medidas de biorremediación y saneamiento de la laguna de Tres Palos, Acapulco, Guerrero*, vol. I, ch. 6, Instituto Mexicano de Tecnología del Agua, Mexico City, pp. 2–42.

World Commission on Environment and Development (The Brundtland Commission) 1987, *Report of the World Commission on Environment and Development: Our common future*, UN, Geneva, viewed 29 May 2007, (http://www.anped.org/media/brundtland-pdf.pdf).

WTO 2004, *Compendium of tourism statistics*, World Tourism Organization, Madrid.

Yokoyama-Kano, A 1981, *La comunidad de aves acuáticas nidificantes de la Laguna de Tres Palos, Guerrero* [*Nesting birds at Tres Palos lagoon, Guerrero*], Tesis de licenciatura en Biología, Facultad de Ciencias, UNAM, Mexico City.

Young, L & Hamshire, J 2000, *Promoting practical sustainability*, Australian Agency for International Development, Canberra.

Websites

iNSnet.com, (http://www.insnet.org). Portal site for sustainable development.

International Institute for Sustainable Development, Canada, (http://www.iisd.org). Policy research institute promoting sustainable development.

Mexican Institute of Statistics, Geography and Informatic (INEGI), (http://www.inegi.gob.mx). Databases and maps can be downloaded from this site.

National Geographic, (http://www.nationalgeographic.com). Contains information on sustainability, sustainable development and geotourism.

University of Reading, Environmental Challenges in Farm Management: Definition, (http://www.ecifm.rdg.ac.uk). Contains several definitions of sustainability.

WorldChanging, (http://www.worldchanging.com). Geotourism is proposed in this website as a redefinition of sustainable tourism.

Notes

1   This fluvial-lagoon system is located in Guerrero state, southern Mexico, and consists of one river, the La Sabana river (Spanish for *savanna*) and the freshwater lagoon of Tres Palos (Three Sticks). The river flows to the lagoon from its springs high up in the Sierra Madre del Sur, in a general south-southeast direction. The river is 57 kilometres long and its basin covers an area of 432 square kilometres (Banderas 1999). Lagoon waters discharge into the Pacific Ocean through a canal and a tidal inlet. Several settlements are located on both banks of the river, but the western margin is heavily populated. The middle section of the river borders the north-eastern part of Acapulco's urban area, where more than 250 000 people live. This study is centred on the Tres Palos lagoon area, but several text references are also made to the La Sabana River and the whole complex itself.

2   Several definitions of sustainability can be found at the website of the University of Reading, United Kingdom. See References listing.

3   One definition of ecotourism is 'responsible travel to natural areas that conserves the environment and improves the well-being of local people' (Honey 1999).

4   The proponent of geotourism is Jonathan Tourtellot, who is based at the National Geographic Society in Washington DC. This organisation has already proclaimed *The Geotourism Charter* that contains the principles in which this new form of tourism is based. Countries can adhere voluntarily to this charter and the first to do so was Honduras, in 2004.

5   There is no consensus as to what ecotourism means. Several authors concerned with the study of tourism, from either a geographical point of view or otherwise, have examined this matter (cf. Hall & Lew 1998; Honey 1999).

6   In Mexico, tourism generates $US8 billion in foreign exchange currency a year (WTO 2004). This is the largest amount of foreign exchange coming from tourism activities in Latin America.

7   At present, the population of Acapulco is estimated at 800 000 inhabitants.

8   There are areas covered with white mangrove and red mangrove (80 per cent of the canal linking the lagoon with the ocean is covered with the former and 20 per cent with the latter). Mangroves on the southern side of the lagoon are in danger as the urban sprawl of Acapulco is demanding more and more land for the construction of hotels, condos and new residential areas. This physical expansion of the city has already reached, at some points, the edge of the Tres Palos lagoon.

9   As requested by the International Geographical Union, at the 2005 Rome Workshop, and with the help of the Tres Palos high school authorities, we selected 10 teenagers, male, aged 16–18 years, resident in the area and enrolled as students at that high school, to carry out the questionnaire survey. One hundred people were interviewed in two of the settlements of the study area: Tres Palos and San Pedro las Playas. These places were chosen due to their geographical location (close to the lagoon), and population size.

10   These characteristics are not really contradictory as one place can be dirty and beautiful at the same time; since 'dirty' is a temporary quality, it can be modified at any time. In view of this, a dirty place that is beautiful may become clean and beautiful if correct measures are put into practice.

9 | Geography for urban sustainable development: Students' proposals to deal with Santiago de Chile urban sprawl[1]

Hugo Romero and Alexis Vásquez

1. Introduction

1.1 Urbanisation in the Southern Cone

In the so-called South American Southern Cone (a region formed by Argentina, Chile and Uruguay, located in the southernmost part of America and surrounded by the Atlantic and Pacific Oceans), nearly 90 per cent of the population is urban—that is, they live in cities and towns. This impressive figure has been reached only in the last 50 years, and in most of the developing countries people are still emigrating from the countryside to urban areas. As a consequence, cities have become, and will be in the future, the most relevant ecosystems in people's lives. However, urban areas concentrate environmental problems such as air, water and soil pollution; natural resources degradation; and severe constraints on quality of living and health. Unfortunately, most of these issues affect the poorest sectors of the urban areas.

Gaining adequate scientific knowledge about the spatial features and processes of urbanisation in developing nations is one of the most important challenges for geographers and the contributions that they can offer sustainable development. If the causes of urbanisation, the nature–society relationships involved, and the effects of these on human health and welfare are well known, it may be possible for geographers to influence policies and programs devoted to the development of the people, and to the protection of environmental and natural resources. Geographers should be active in designing, implementing and assessing proposals about sustainable development in urban areas, and formulate management plans and policies that can ensure a better quality of life for present-day and future city-dwellers.

One of the most relevant causes of degradation of natural resources and the occurrence of natural hazards is that urbanisation has occurred in a fast and homogeneous manner, unlike the previous heterogeneous mosaic formed by prime agricultural lands, floodplains and natural ecosystems. The size, complexity and speed of urbanisation are related to socioeconomic changes that have taken place on global, national and regional scales.

In Latin America, the urbanisation process has accelerated since 1950 because all the countries followed a socioeconomic program called 'Import Industrial Substitution'. This program favoured industrialisation as the only way that Latin American countries could reach sustainable economic development, supply domestic markets with necessary goods and services, and become independent of international markets. The industrialisation efforts brought manufacturing plants to the cities and attracted large populations from rural areas, where there were high unemployment rates and low incomes in the agricultural sectors.

The small size and economic limitations of domestic markets—a consequence of prevalent poverty, and the political impossibility of integrating all Latin American countries—meant the failure and gradual abandonment of industrialisation processes, increasing the number of unemployed and poor people. From 1960 to 1970 there occurred an intensive process of social polarisation in Latin American countries, with conflicts occurring between the skilled labourers who worked to supply urban services and markets, and the unskilled and unemployed who remained marginal inhabitants. This general socioeconomic division of people strongly influences the growth and structure of urban spaces.

Higher income groups have abandoned central areas and migrated to peripheral zones, looking for socially segregated areas. Specific geographical areas were selected to receive the middle and upper classes, following in many cases the design and pattern of North American cities (low-density, open spaces, green areas, good connectivity and private transportation). Meanwhile, the impoverished sectors occupied unsuitable zones, far away from central city services, self-building precarious homes and slums. Because of the sudden appearance of these poor urban settlements in rural areas or outside the borders of the cities, they were called 'mushroom slums' in Chile and 'misery villages' in Argentina. The new urban development areas of the rich, and the slums of the poor urban population, meant there was an impulse for urban sprawl.

Social polarisation and the end of the industrial period gave rise to political troubles for the Southern Cone, and as a consequence military governments were installed, consolidating the social segregation process. In 1979, in Chile, the military government declared an absolute freedom to urbanise the land and replaced public regulations that controlled urban sprawl with market-oriented private initiatives. Land privatisation and commodification aggregated immense areas of rural land with cities. Additionally, for political and economic reasons the government accentuated social segregation by moving the poor settlements that were located between middle- and upper-class areas to outside sites. The spatial concentration of poor people in specifically designated locations not only allowed political control but also increased the market value of previously occupied lands.

Since 1980, Latin American cities have felt the impact of globalisation on the economy. They have experienced the rapid development of telecommunications, and the increasing use of private cars as one of the main transportation systems. Large cities have been converted into city-regions, or nodes of complex national and international spatial networks. Labour, goods, services, technologies and communications must now be supplied for extensive regions that exploit outward-oriented natural resources and raw materials for global markets. City-regions extend geographically at increasing distances, covering previously natural and rural landscapes with homes, malls, roads, highways and parking areas.

The social segregation process has been reinforced by globalisation. Members of higher income groups who work for transnationals and national-related companies that supply goods and services for them, have received direct benefits and acquired new large and low-density homes beyond the city boundaries. City sprawl is now progressing along roads and highways to the outskirts of cities, and taking rural lands in a 'frog jump' pattern. Isolated and unconnected urban areas appear in the middle of agricultural lands, requiring an enhancement of urban services and facilities at higher social, economic and environmental costs.

The price of urban land has substantially increased as a result of the demand for its use. As a consequence, the poorest people have been obliged to build or purchase their homes in areas located far from the city centres and richer areas. State deregulation and privatisation of the land have been important factors in the unlimited sprawl of Latin American cities during the last few decades.

1.2 Effects of urban sprawl

From the point of view of physical geography, most of the Argentinean and Chilean Andean cities are located in the centre of large watersheds. In earlier times, between 1500 and 1700, and following strict Spanish ordinances, the location of cities was determined by water availability and drainage, good defence capabilities against the indigenous population, and suitable agricultural lands for food supply. Watersheds are complex, integrated environmental systems and any land use or cover change that takes place anywhere inside the area affects the performance of the whole area. Highland–lowland interactions are critical to maintain dynamic balances of energy, materials and information between mountain slopes and floodplains. Deforestation, soil erosion and water pollution have produced considerable environmental damage in the watershed areas. Recently, watershed urbanisation has negatively affected environmental health, especially through the imperviousness of soils, and air and water pollution (Barnes et al. 2001).

City sprawl causes severe negative effects on natural and social urban environments. Urbanisation means that soils are covered with solid surfaces, affecting water, energy and biological cycles. Ground infiltration of storm flows and deep storage of water is replaced by large and instantaneous superficial run-off after rain. Cold vegetated areas are replaced by heat islands formed by impervious covers, increasing the climate-warming process. Green areas, their shadows and their contribution to atmospheric humidity through evapo-transpiration are eliminated. Photosynthesis and gas recycling are also diminished, increasing the greenhouse effect and air pollution (Romero et al. 1999; Romero et al. 2003; Romero & Vásquez 2005b).

Urban sprawl also increases the social and physical distances between city inhabitants. This requires the installation of urban facilities at growing distances from city centres, at higher economic and environmental costs. Paradoxically, central services and infrastructures are underused in many cities of developing countries. Longer distances and travel times are then required to move between home and the workplace. As a consequence, traffic jams and air pollution caused by mobile sources have increased, negatively affecting public health. Large agricultural lands are being continually lost. Valuable soils have been paved, causing increasing surface run-off and flood hazards. Vegetation patches and corridors have disappeared or are being fragmented, and their environmental services (cleaning and moistening the air, mitigating urban heat islands and facilitating water infiltration) have been reduced or eliminated.

1.3 Study areas: Peñalolén commune and Pudahuel municipality, Santiago

Santiago, the capital city of Chile, has nearly six million inhabitants (INE 2002) and, like Buenos Aires in Argentina and Montevideo in Uruguay, concentrates nearly 50 per cent of the national population and more than 60 per cent of the economic activities and products. As a result, these cities also concentrate most of the urban environmental issues. Huge spatial concentrations of economic activities and a crowded population in a small space threaten local ecosystems and their components—water, soil, climate, flora and fauna—and increase pollutant emissions and wastes.

Two areas of Santiago affected by recent urban sprawl have been selected for geographical teaching. They are located in opposite geographical, natural and social areas of the city.

Study area 1: Peñalolén commune
The first area corresponds to the Peñalolén commune, located in the Andean mountain foothills that surround the eastern border of Santiago (see Fig 9.1, col. sect. opp. p. 224). Upper- and middle-income social groups who try to escape from the air and water pollution of central areas are currently urbanising this area. They build gated villages, looking for good environmental conditions, social homogeneity and security.

However, these new developments cannot avoid the numerous natural hazards occurring in the sensitive environments that they occupy. Foothills are an interface between mountain slopes and floodplains, and are often drained by several creeks, streams and rivers. Mediterranean-type shrubs and bushes on the slopes, and vines, grasslands and some native forests on the floodplains, originally covered these pre-Andean landscapes. These vegetation covers ensured the infiltration of winter storm flows of waters and supported the feeding of groundwater sources. Groundwater storage is fundamental for water supply in the arid lands where the city is located. The long-term annual average rainfall is only about 300 millimetres and occurs in just 20 days from May to August (the southern hemisphere winter season) (Romero 1985). For the rest of the year—about 240 days on average—the city depends on water from rivers, groundwater, snow and glaciers.

Given the fact that this foothill area was previously completely marginal for city growth, it was mainly considered by urban planners and regulators as an ecological protection zone, occasionally and illegally inhabited by poor people. The new upper and middle classes moving towards the foothills are, in some ways, invading natural landscapes and social spaces, and there will be serious environmental and sociocultural collisions between natural and urban landscapes, and between different social groups, if urban spaces are not well managed.

Santiago city is divided into 36 different communes, and each is under a separate municipal government. Six of them are in charge of the Andean foothill area. The absence of a unique government for the whole city, and the lack of coordination among numerous sectors and local governments, imposes institutional constraints that affect environmental management. The selected study area in the eastern section of the city is under the governance of the Peñalolén municipality.

Study area 2: Pudahuel municipality

The second study area is located on the alluvial plains and foothills of the coastal range close to the watershed where the city is situated, on the western and lowest side (Figure 9.1, colour section, opp. p. 224). Given the fact that these lower areas accumulate wastes, and there is air, soil and water pollution, these land values have always been the lowest in the city. As a consequence, there have been extensive state housing programs to accommodate the poor. Chile's principal international airport is also located in this low-valued area.

These rather marginal areas have, however, become relevant for the city-region. New roads and highways have increased their connectivity with the rest of the city, the region and the country. Improved connectivity has in turn attracted richer dwellers, and numerous industries and services related to the production of goods and services for export. As in the first study area, urbanisation of unsuitable areas and social segregation are producing related geographical issues.

This western side of the city is under the governance of four main municipalities (Maipú, Lo Prado, Pudahuel and Lampa), but the study area belongs specifically to the Pudahuel municipality. Nearly 150 000 dwellers of low and medium income traditionally lived in this area, but as a consequence of the urban sprawl, there is pressure to urbanise agricultural areas and grasslands. The main geographical issue is how to reduce the urban occupancy of an area that is already environmentally polluted and degraded, and how to ensure natural conservation, the restoration of polluted environments and social integration.

2. The students, their teaching materials and their aims

2.1 The student projects

The aim of the first exercise was to produce an Environmental Management Plan for the Peñalolén Municipality, located in the Andean foothills. This area was severely affected by floods during June 2005, when nearly 150 millimetres of rain fell in only a few hours. Numerous urban areas were flooded, and run-off, erosion and landslides destroyed many roads. It was evident to the citizens that urban sprawl over unsuitable land has increased flood hazards in impervious areas, and reduced infiltration zones. Specific municipal proposals and programs must be prepared to control foothills urbanisation, control impervious areas, restore vegetation cover and protect downstream areas, especially those areas occupied mainly by low-income families.

Members of the Environmental Planning and Management Masters Program, supported by the Department of Geography at the University of Chile, and within the course area of Environmental Integrated Management, were invited to participate in this exercise. They were professionals, each around 25 years of age, from different disciplines such as engineering, forestry, chemistry, biology, architecture and geography, who were interested in municipal environmental management. This interdisciplinary group of students studied three different environmental components: physical components, socioeconomic components and the built environment. Additionally, institutional issues were specifically considered because they are fundamental for environmental management on a local scale.

The management component contained two parts. The first was an environmental diagnostic that concluded by identifying the main integrated group of issues. The second part included some sets of short- and medium-term action proposals that could be taken into consideration by local authorities to manage the current problems. Proposed actions were aggregated and presented to local authorities such as the Municipal Council, land use planners, and socioeconomic development agencies. From a selection of aims, issues and actions, the proposed plan was discussed with local community representatives. As a consequence, they were adjusted and modified, then submitted to the municipality's Mayoral Office.

In the case of the Pudahuel municipality, participants were first-year geography students at the University of Chile. Their ages were around 18 years and for most it was the first time that they had received information about the aims, content and values of geography and about its ability to analyse and solve some ongoing issues of urban sustainable development. Most of the first-year geography students in the university had not had previous geographical education in secondary school. Some only had tedious lessons about irrelevant issues such as the naming of capital cities around the world. In Chile, geography has always been a minor subject allied with the teaching of history and social sciences.

These are not the only reasons for the cultural underestimation of geography in Latin America, but they constitute a large handicap when recruiting motivated students at university level and, in turn, educating good professionals to produce useful knowledge. Another reason is the lack of relevant and updated scientific and applied research. This is especially the case in Chile, one of the longest countries in the word, where no more than 20 researchers have to study large and complex geographical transformations from the dry Atacama Desert located in the north, to Mediterranean landscapes in the centre, and template rainforest, lakes, volcanoes and ice fields in the south. The highest Andean summit (above 6000 metres) and the world's longest and deepest shoreline are only about 200 kilometres apart.

2.2 Chile's socioeconomic and geographical background

Economically, Chile is currently the main copper exporter in the world and satisfies nearly 45 per cent of world demand. It is also a large producer of iron, molybdenum and gold, and an important producer and exporter of wine, salmon, fishmeal, subtropical fruits and vegetables, wood, pulp and cellulose.

Large Chilean geographical transformations are related to its rapid economic, social and cultural growth. Chile has increased its gross domestic product and its per capita income by 150 per cent during the last 20 years. Economic growth reached an annual average of 5 per cent between 1983 and 2005, and has allowed a 50 per cent reduction of poverty levels. Socioeconomically, Chile's successful progress, based on free markets, and state reduction, deregulation and privatisation of goods and services, is presented as a development model.

Beyond the political interpretation of these facts, it is important to recognise the geographical complexity of these transformations. A good performance on a global scale does not necessarily mean success on regional or local scales. How to increase the local and regional benefits from globalisation is one main challenge for Chilean geographers. Regional disparities and social segregation have increased as a consequence of this economic success. Social and economic differences have increased among regions, between modern and traditional farmers, between large and small industries, between large and small cities and between richer and poorer classes.

Traditional social services provided by the state, like public education, housing, social security and health, have been privatised and are unevenly distributed in the national population. Social inequality is threatening the achievement of real sustainable development.

Recent Chilean socioeconomic development has produced some negative environmental outcomes. Mining, fishing, forestry and aquaculture have produced immense environmental degradation in some previously pristine areas. Air, water and soil pollution can be found everywhere in rural and urban areas. Privatisation or commodification of space, air, land, soil, water and many natural resources has produced serious problems in terms of social accessibility and appropriateness. The privatisation of land, resources and spaces has been constitutionally granted, rejecting as a consequence communal property systems. Traditional communal goods and services have been threatened or have disappeared, affecting urban, agricultural and indigenous communities.

Chile is the only country in the world where water rights are considered a free-market good. Throughout the country, and especially in arid lands, mining, water and development companies have obtained permanent water rights by purchasing them from poor indigenous and rural communities.

2.3 The project challenges

Chile's recent economic development and its lack of sustainable development is common to most Latin American countries, and constitutes a formidable challenge for young geographers.

Spatial and environmental transformations in Chile seem to explain why, in the last few years, an increasing number of students have obtained high marks in national selection tests to enter the public universities and are choosing geography. It is hoped that these new geographers will help address the lack of geographical knowledge in the country. The role of new technologies to acquire and capture geographical information is an important part in this education process. The teaching of subjects such as Geographical Information Systems (GIS) is a priority in developing countries. Access to free high-resolution satellite images, like GoogleEarth, and the low cost of access to information, data and computational systems, are important parts of their teaching and learning.

The scientific, theoretical and conceptual basis of the discipline is as important as technological topics in its introductory courses. In the University of Chile's first-year geography program, lectures about the ability of geography to solve relevant sustainable development problems and environmental issues at local and regional levels are important. The course program includes fundamentals of regional geography, emphasising concepts like differentiation and classification, diversity and complexity of natural and cultural spaces, and the allocation of land use according to land capability and suitability.

A simple manual of geographical information systems was created by each of the 20 groups within the first-year class. Each group comprised three or four students. The general aim of the exercise was to introduce some concepts and methods to analyse and solve selected issues, preparing small research pieces on relevant environmental problems such as the environmental impacts of the urban sprawl in Santiago city. Other topics included urban suitability of the coastal-range foothills located at the western side of the city, the proposal to locate a recreational park to protect nature conservation areas, and the environmental impacts that could be caused by the expansion and growth of the international airport on surrounding areas.

A geographical information grid was created from 1:50 000 toposheets, and basic statistics of some geographical features were calculated, such as slopes, aspect, drainage length and density, roads and pathways, land use and land cover. Satellite images and aerial photographs were also used for spatial analysis as well as introductory fieldwork. Spatial covariance and correlations were analysed and statistics were complemented by interviews with local communities during organised fieldwork.

As a general outcome, students were much more interested in geography at the end of the semester. They participated in a process of 'learning by doing' that included cartographical analyses, image and aerial photograph processing, and fieldwork and data analyses. They prepared in the classroom the materials and information to be used in the field and selected sampling areas. Finally, they presented and discussed the results to the class and talked about them with local communities. When they confronted some of the most urgent environmental issues in the city where they live, and then gradually internalised scientific methods and concepts, they were much more aware of how they could improve their social contribution as environmentally committed citizens.

3. Evaluating environmental sensitivity and proposing a municipal management program for the environmentally sustainable development of the Peñalolén Andean foothills

The aim of the first study, involving students from the Masters program, was to use environmental diagnostics to propose an Integrated Management Plan to municipal authorities and local communities which allows for sustainable communal development. Such proposals have to take into consideration political definitions and sources of funds, and assessment of conflicts, uncertainties and risks related to their implementation, before their aims and targets can be programmed.

The environmental diagnostics include the physical, socioeconomic, urban and institutional components. For each of these, selected variables were analysed and integrated to determine how they are affected by urban sprawl.

3.1 The physical environment

The Andean foothills are composed of very steep slopes on the highlands, above 1200 metres, which correspond to pre-Andean mountain areas, and a lower but less steep piedmont, located between 600 metres and 1200 metres, where alluvial and colluvial fans interact between the mountains and the floodplains along several creeks and streams (Table 9.1). Taking into consideration flood hazards represented by the storm flows that descend from the upper zones along these drainage systems, some riparian buffer zones of protection against floods have been considered (Zandbergen et al. 2000). They cover a 200-metre wide corridor along the borders of the main streams and the main irrigation canal that cross over the commune, and 150 metres and 100 metres wide in other areas according to the discharge of streams and creeks.

Table 9.1   The main creeks and streams in the Peñalolén area

Creek name	Watershed size (km²)	Number of tributaries	Altitude (m)
Peñalolén	1.93	7	2340
Nido de Aguila	4.11	8	2552
Lo Hermida	1.05	4	1100
Macul	12.03	12	3253

Source: Aravena (1994)

As an indication of the lack of consideration given to floods, all the main streams and creeks that originate in the highlands flow towards the urban areas along paved roads and avenues: Lo Hermida stream continues down to Grecia Avenue; Nido de Aguila stream forms José Arrieta Avenue; and Peñalolén Creek forms Talinay Avenue.

The uppermost zone is completely uncovered because of the arid and cold climate that dominates the Central Andes. On the upper piedmont proper, shrubs cover most of the slopes and some native forests occupy the southern aspect mountain slopes. Cross-tabulating altitude and aspect, the dominant vegetation of the southern exposures between 720 metres and 1000 metres is *Acacia caven*. On the drier northern-faced slopes, between 800 metres and 1000 metres, the main species are *Quillaja saponaria*, *Baccharis linearis* and *Acacia caven*. On humid southern slopes there are forests containing species such as *Quillaja saponaria*,

Cryptocarva alba and *Lithraea cáustica* (CONAF–CONAMA 1999). At the lower ecological belts, some forests and shrubs still exist along the streams and between some remaining grasslands and crops. In the central area of the commune, a large vineyard and a neighbouring green area are still surviving in the middle of the urban land. However, the sustainability of these land covers is uncertain since most of this area is currently being built upon.

Land cover is shown on one map of the general distribution of vegetation, indicating the presence of three main zones: natural vegetation at the higher lands, some crops and vineyards in the middle areas, and the almost complete lack of vegetation in the lower zone where urbanisation is currently taking place. High productivity could be found only along some creeks and streams and remnants of agricultural lands, but for most of the community, vegetation productivity is low to very low.

The most environmentally sensitive areas, which require special conservation measures, are small patches (3 per cent of the total area) of native forests and shrubs located on the southern aspect of creeks and streams that drain the upper zone. Extensive areas of high sensitivity (29 per cent of the total surface) are located on the Andean slopes and along the borders of larger streams and irrigation canals. Only some green areas and cultivated land located in specific patches in the upper zone and as remnants of farms and vineyards still remain in the urbanised area.

3.2 The socioeconomic environment

For the socioeconomic environment, several basic variables and indicators were considered, such as housing quality, population density and the percentage of people living under the poverty threshold in each of the vicinities.

Waste deposits

Housing quality includes the spatial proximity of the population to micro deposits of domestic wastes (Figure 9.2) (I. Municipalidad de Peñalolén 2001). The illegal deposit of domestic waste in abandoned lots of land inside the urban areas, along rivers and stream floodplains and in any site that is not occupied is a common threat to public health in all Latin American urban areas. These micro areas are important sources of zoonosis and infection that require permanent control by municipal authorities.

There are nine vicinal or neighbouring units that have high risks for the public health of their dwellers because of their proximity to sites where domestic wastes are illegally deposited along stream beds, abandoned lands and irrigation canals. Unfortunately, poor people inhabit all of these areas. As a contrast, middle- and upper-class dwelling areas have medium or lower risks because they are located far away from the deposit areas, and have a good municipal collection system or their own landfills.

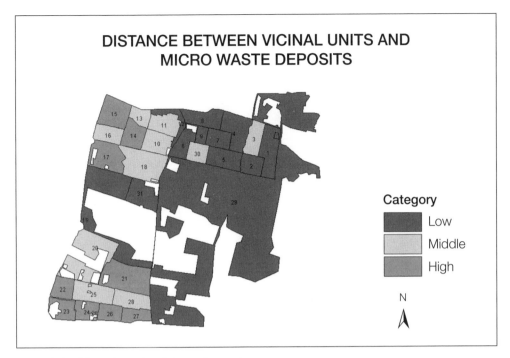

Figure 9.2   Micro deposits of domestic wastes

Crowding level

Another interesting socioeconomic indicator that has been taken into consideration is the 'crowding level', corresponding to the number of inhabitants per room of each home (Figure 9.3, page 158). The highest number of people per room is in Vicinal Unit No. 8 in the northern border of the commune, which corresponds to an illegal occupancy of land that occurred five years ago. This kind of illegal urban occupation and the formation of slums in some parts of cities by very poor people is exceptional in Chile today but it was very common 20 years ago, and unfortunately is still very common in several Latin American cities. Hundreds of poor families select a piece of abandoned or empty land and build very small and precarious homes, mainly made from waste materials ('mushroom constructions'), where families fight to survive, to build a home, to get basic facilities and services from governments, and lately, to be moved to more suitable zones and better housing. In comparison to the precarious situation of Vicinal Unit No. 8, the rest of the poorer areas have an intermediate level of crowding, and middle- and upper-class areas present a low to very low level (INE 2002).

Socioeconomic differences and social segregation even between short distances in urban areas are main characteristics of the current state of urbanisation in Latin America. Several indicators could be used to demonstrate such differences. In the

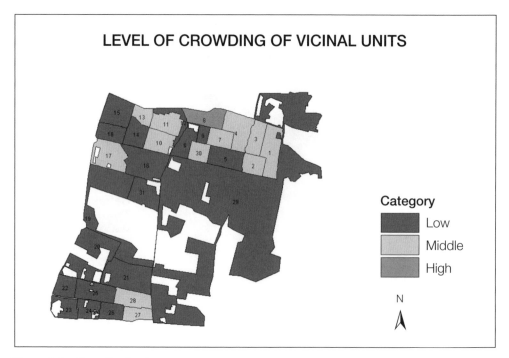

Figure 9.3   Crowding level

following paragraphs, the access to sanitary facilities, poverty percentages, dwelling densities and the quality or state of homes are presented as useful geographical indicators to deal with social segregation.

Access to basic facilities

The accessibility to basic facilities (drinkable water and sewerage systems, Figure 9.4) indicates again that Unit No. 8 is lacking both services as a consequence of its illegal land occupation. Vicinal Unit No. 1 also demonstrates these limitations, which is rather rare in a commune where most of the population has good or regular access to networks of privatised facilities (I. Municipalidad de Peñalolén 2001).

Distribution of poor people

The complexity of the socioeconomic mosaic of the Peñalolén commune can be observed from the distribution of the relative numbers of poor people in neighbourhood units, as shown in Figure 9.5 (INE 2002).

The map shows that the population in Peñalolén represents the three main Chilean socioeconomic groups: upper, middle and lower classes. This is an exceptional situation for Santiago city and other Latin American capitals at present, since each commune tends to be an exclusive residential or commercial area for specific social classes. The reasons for this rather exceptional social mixture are

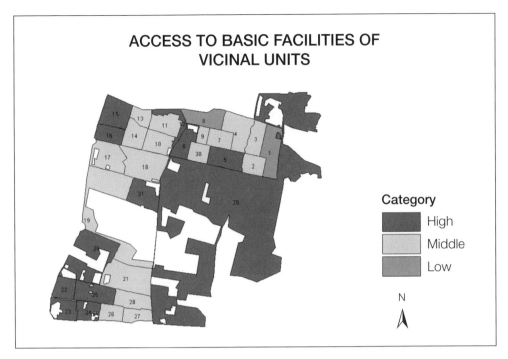

Figure 9.4   Access to basic facilities

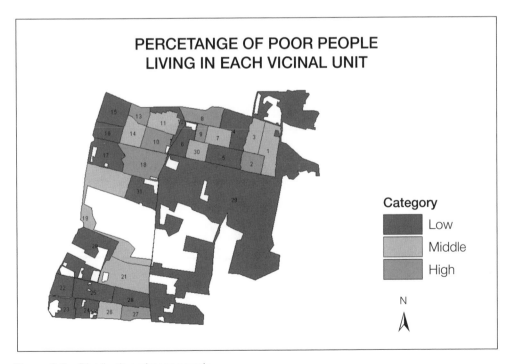

Figure 9.5   Distribution of poor people

related to the historical arrival of people. The inhabitants who arrived before 1970 were mainly poor and middle-class people looking for marginal and lower price lots of land that currently remain in urban consolidated neighbourhoods, like Vicinal Units 1, 7, 10, 13, 14 and 19. Middle-class groups who arrived between 1970 and 1990 were attracted by small parcels of land on which to build a house in a lower density area with good environmental quality. Recent inhabitants, who have arrived since 1990, belong exclusively to the upper and middle classes searching for better environmental quality and isolated lots. During the last few years, younger upper middle-class families have arrived to live in gated villages. During this decade, some illegal occupation explains the presence of poorer areas such as Vicinal Unit No. 8 (I. Municipalidad de Peñalolén 2001).

Poor people are defined in Chile as those urban inhabitants whose income is not sufficient to satisfy basic needs such as food, housing, clothing, and services such as education and health. Figure 9.6 shows that higher percentages of poor people are spatially concentrated in small areas called 'poverty pockets'.

Lower urban densities are other indicators of affluence in the urban environment. There is an inverse spatial correlation between urban population density and the income of different socioeconomic groups. Most of the areas where poor people live correspond to high-density urban zones. Conversely, most of the areas with lower percentages of poor people are also areas of lower residential density.

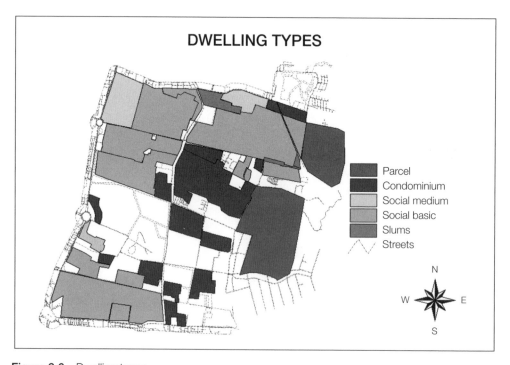

Figure 9.6   Dwelling types

Most of the observers of the socioeconomic conditions in the urban areas realise that there is a direct geographical correlation between the level of family income and the state of conservation of dwellings. Clearly, if family income is not enough to satisfy basic needs, then house restoration is unlikely, so there are meaningful differences in the presentation of homes in terms of type and the state of walls, roof material, painting care, presence of individual gardens, number of paved roads, urban equipment and so on. These differences determine in turn the relevant environmental differences in terms of soil or air pollution, green areas and security, forming complex and integrated spatial patterns that transform social segregation into environmental segregation. The different Chilean urban environments represent social differences in a clear manner.

Synthesising the socioeconomic variables and indicators discussed above shows that one vicinal unit, No. 8, is classified as an area with a very high sensitivity that requires urgent measures. The municipality has already made the decision to remove this population to safer housing. Three other vicinal units (7, 10 and 17) also need attention from social institutions and programs.

3.3 The urban environment

A final set of geographical analyses focuses particularly on urban environmental issues and challenges. Land use and land cover fragmentation which results from haphazard urban occupancy can result in urban areas coexisting beside agricultural, industrial, commercial and bare lands. There are no central services located to minimise the distance and to spatially integrate different segments of the population. Urban fragmentation can be seen in Figure 9.6, which represents different kinds of residences: condominiums, parcels of agricultural land, basic and medium social housing, and slums. See also Figures 9.7 and 9.8, colour section, opp. p. 224. The foothills' topography, the increasingly steeper slopes upstream, the lack of centrality and the social fragmentation that characterise the urban environment result in lack of connectivity. There are no circular or ring roads to facilitate connectivity between dispersed services and the spread and fragmentation of residential areas. Traffic jams are a current and future inevitable urban and environmental issue that results from the lack of effective urban planning.

On an urban scale, the presence, abundance and state of the green areas is another great challenge for environmental planning. Accessibility by the local population to vegetated areas is a significant issue in developing countries, especially considering the privatisation and commodification process that has taken place in recent years. This process has been a consequence of the application of strict neoliberal economic models and the diminution of the role of the state in facilitating social welfare (Romero & Vásquez 2005a). There are no open green areas or public parks in this commune. Only some small patches of vegetated cover exist along the buffer zones of the main irrigation canal. Cultivated areas are private properties that cannot be used or accessed by the community. Upper creeks and streams, and wood forests, are also private land with restricted access for the population.

3.4 The flood hazards

Three main factors explain the high-risk nature of flood hazards for the community. Firstly, there is the commune's location in the foothills, where many streams and creeks cross the land. Many of these sporadic streams are activated by storm flows during winter rainfalls, and they can overflow to older floodplains where houses have been built (Corvalán et al. 1997) (Figures 9.9 and 9.10, colour section, opp. p. 224). Secondly, the community is crossed by two large irrigation canals, which are used like drainage systems for streams and creeks. Their discharge capacity could be exceeded during storms, flooding the neighbouring downstream areas. A third reason is the increasing imperviousness of the ground, which results from constant and progressive urbanisation. Land use and land cover changes caused by urbanisation mean increased run-off and decreased natural soil infiltration capacity (Arnold & Gibbons 1996; Barnes et al. 2001).

Based on local knowledge about flooding and waterlogging after rain, and taking into consideration the vulnerability of the urban areas, communal sensitivity to flooding can be analysed. The highest sensitivity is observed when some streams are redirected and continue the storm flow along (under or above) paved roads that connect the foothills with the floodplains, or when they intercept irrigation canals. These canals cause floods when they cross a large part of a populated area. Maintenance of canals and the implementation of vegetated riparian buffer zones are measures that should be taken by local communities and authorities. Low sensitivity to floods occurs in the main slope breaks and some areas that suffer waterlogging after rain.

3.5 Integrated environmental sensitivity

Urban sprawl is a severe environmental issue for these areas. Deforestation and the increase in impervious areas transform the hydrological cycle at a local scale. Water infiltration in the ground is replaced by superficial run-off, increasing the occurrence of floods (Romero & Vásquez 2005b).

Another relevant issue is related to the scarcity of green areas inside the community. Green open spaces facilitate social integration, and supply important environmental services and functions such as air and water cleaning, regulation of urban climate, erosion control and wildlife habitat protection. Also, sensitive areas tend to generate brown fields which are used as illegal waste deposits, which is one of the most critical environmental issues in poorer residential areas.

3.6 The Institutional System for Environmental Municipal Management

Creating environmental municipal unity

The Chilean Political Constitution is the most important piece of legislation for national institutions. The Constitution establishes that the state must offer all Chilean citizens the right to live in a healthy and good environment. To reach this aim, constitutional law, like municipal law, has to interpret this principle. Accordingly, Chilean municipalities are in charge of environmental protection at a local level, and to achieve this goal they can act directly or indirectly, implementing their own instruments and ordinances in the former case, or collaborating with the institutions that have legal responsibilities for the care of the environment in the latter case (Espinoza et al. 2000).

However, according to the National Environmental Commission, only 19 of the 310 Chilean municipalities have created some environmental municipal unity. Of these:

- 39 per cent, especially in small communes, depend on some Community Development direction
- 19 per cent depend on the Communal Planning Office
- 16 per cent depend on the Waste Treatment Department
- 15 per cent depend on the Director of Works in charge of construction permission
- 11 per cent use other organisations.

The institutional system and organisation has not been a central focus in the development of environmental municipal management plans and procedures in Latin America. Most of the time, analysis of environmental management has focused more on institutional intervention in the components and interrelationships within ecosystems, rather than a concern for the political, social and organisational components of the municipal governments.

Municipal management plans and models

To define and discuss the environmental management institutional system, the students working on this study based their analyses on the aims and implementation measures included in the *Communal Plan of Development 2001–2005*. This is an official report, where municipal authorities set out their goals, targets and alternative procedures to implement sustainable development. Unfortunately, these analyses took place when the current plan was ending, and the authorities' interest was on drafting a new one. The time of validity of this planning instrument, though, is strictly related to the period of government of municipal authorities. In 2005 a new municipal government with responsibility for the Peñalolén area had taken over, and it did not necessarily recognise the value of the instruments and initiatives developed by previous authorities. Short-term validity and political dependence of the municipal plans is a large restriction for long-term sustainable development plans for Chilean communities.

Other considered sources of information were the acts of the ordinary and extraordinary sessions of the Municipal Council recorded between November 2004 and November 2005, including the municipal administrative ordinances, the assessment of municipal performance indicators and several interviews with municipal professionals. The aims of the analyses of the recorded information were to characterise the current model of municipal management, especially in terms of its structure and functions; to have an idea of the relative importance of the environmental issues in respect to the general problems that concern the municipality; and to understand the main information pathways and procedures inside the municipal structure.

A complete model of municipal management is presented in Figure 9.11. This model shows a sequence of operations, starting from *strategic fundamentals* represented by the institutional mission and vision. This phase is followed by the *planning* steps where the strategic axes are defined, and the alternatives, aims and

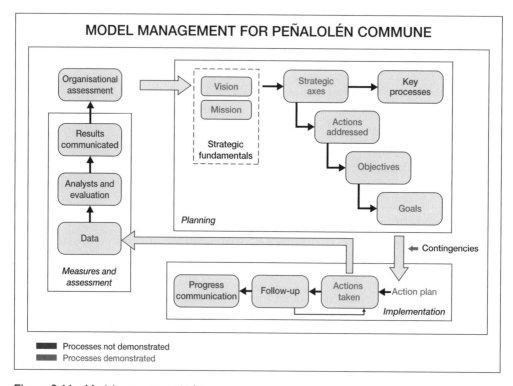

Figure 9.11   Model management plan

methods are developed. The *implementation* stage includes decision-making and actions taken, follow-up of results, and the communication of progress. Finally, the model ends with the *measures and assessment* process, where the organisation collects, analyses and evaluates data, and finally communicates to assess the organisation's performance. It is important to know that only some of the steps and procedures have been developed by the current institutions, and that the main gaps are caused by the lack of follow-up procedures and communication.

A second model of municipal environment management that was considered contains the hierarchical functions that link the *strategic level* with the *operational level* of functioning (Figure 9.12). While the strategic level seems to be well-developed and implemented, there are no clear relationships between this level and the operational organisations. The lack of linkages between both levels can even mean that they operate in opposition, contradicting strategic missions, visions and decision-making. Furthermore, operational decisions would not act as feedback for the correction and adjustment of the institutional model (Espinoza et al. 2000).

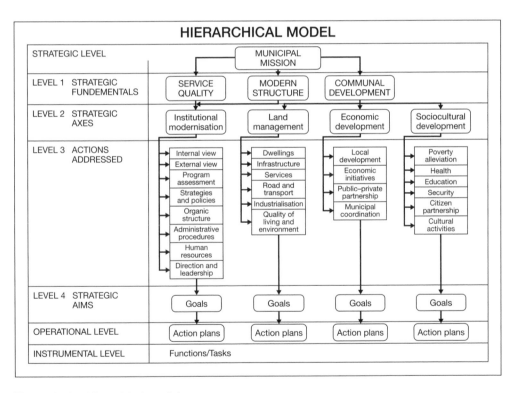

Figure 9.12   Hierarchical model

As a consequence of this organisational weakness, there are differences, omissions and even contradictions between the environmental mission, vision and goals of the strategic levels and the real decisions that are taken at operational levels.

Some strategic environmental aims included:

- developing awareness in the local population of environmental concerns
- the reduction of environmental and indoor pollution concentrations
- formal and informal environmental education to local communities
- promotion of the use of non-contaminant energy sources
- conservation, protection, maintenance and restoration of the communal natural resource base
- the facilitation of access for local people to the Andean foothills and piedmonts
- the implementation of nature reserves
- the promotion of domestic waste recycling.

However, thanks to external initiatives, two of the goals have been partially implemented. In two municipal primary schools, the teaching of environmental education has been included in the basic cycle of teaching from first to fourth year. In terms of the conservation of natural resources, the municipality has decided on the creation of an urban park and the forestation of slopes in a specific area of the piedmont. The implementation of the rest of the initiatives remains without a clear institutional responsibility, assuming that, in the absence of a proper environmental institution, they should be managed by municipal departments in charge of waste, green areas or communal health.

Given the high number of illegal domestic waste deposit sites, the necessary integral treatment of domestic wastes is a good example of institutional requirements that are not fully met in the current situation. This program requires the simultaneous participation of different municipal offices and departments, including formal and informal education, in-home separate collection of items and recycling, and regular follow-up, enforcement and assessment. The institutional coordination of integrated systems of management does not exist in the current structure.

Environmental Municipal Ordinances

A second group of institutional instruments to manage the environment is formed by domestic regulations that are called Environmental Municipal Ordinances. Peñalolén Municipality generated 18 ordinances during its last term in office. Six of them regulate environmental issues such as domestic waste treatment, harmful noise, sanitation, performance of open markets and fairs, the urban master plan, and regulations about green and forested areas that should follow the new urban developments in terms of landscape architecture.

However, there are many difficulties which explains why there is only partial enforcement of some of these Ordinances, including:

- the spread of control and administrative procedures in many non-coordinated departments
- the lack of personnel, training and financial support

- the current level of education
- behaviour of the population in terms of noise or waste.

Although there is no evaluation of the institutional performance, the existence of 55 illegal waste deposits is the best evidence of the limitations that exist for environmental management.

Given the problems that could be observed in this system of municipal environmental management, the municipality is trying to modify the current organisational structure by incorporating a specific unit in charge of these issues which can coordinate decision-making and implementation between departments and offices, and can support the Municipal Council and Mayor's Office functions. A geographer has been selected as head of this new unit. A second set of measures is related to the implementation of symbolic public works, such as the construction of a municipal park, access to public footpaths, transformation of waste deposit sites into recreational areas, and development of new green open spaces that can strongly improve the environmental perception and concern of the local people. Figure 9.13 presents the proposed new municipal structure.

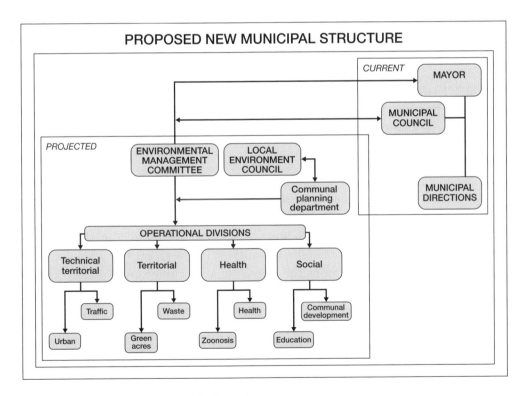

Figure 9.13   Proposed new municipal structure

- A municipal *Environmental Management Committee* that reports to the Mayor's Office and the Municipal Council has been established. This Committee must coordinate and ensure environmental considerations are included in the decision-making processes about urban development, traffic control, green areas, waste treatment, public health, education and community participation. The Committee must produce regular reports about municipal management performance, and monitor and follow up the achievement of environmental targets.

- A second committee is called the *Local Environment Council* and is composed of sector and community representatives nominated by the Mayor's Office. This is mainly a communication body that allows the municipal officers to know about, and solve directly, the environmental requirements and issues identified by the local population.

- The geographer who will coordinate both the Environmental Management Committee and the Local Environment Council has to plan the municipal strategic lines of action, assess ideas for environmentally sustainable development, set up and coordinate environmental actions inside and outside the municipality, and coordinate and control studies and research that are needed to improve environmental management.

The student group prepared a list (Table 9.2) of initiatives and projects that the new environmental office has to propose to the local government and to the communities.

4. Land use and environmental changes and challenges caused by Santiago's urban sprawl on the western-side coastal range (Pudahuel commune)

4.1 Looking for suitable urban areas

As mentioned previously, Santiago de Chile is situated in a closed watershed, surrounded on the eastern side by the high Andean mountain chain. From these mountains, a large river (Mapocho), and a set of large and small streams transport water and sediments, often torrentially, towards the Pacific Ocean just 250 kilometres from the Andean summits. Among the largest streams, Lampa and Polpaico are the most important. The concentration of drainage in this part of the city means that the toponomy Pudahuel has given to this commune maintains the indigenous meaning of the term—'the place where waters converge'.

Steep slopes and a lack of vegetation to protect the soils are the main cause of soil erosion and natural hazards such as mass slide movements of sediments. The semi-arid condition is another meaningful geographical feature. In Santiago the annual average rainfall is only 300 millimetres from just 20 days in a year, concentrated between May and August (the winter months). Large annual variability is another relevant climate feature. During El Niño years, rainfall can reach 800

Table 9.2 Proposed initiatives and projects of the new environmental office

Scope of the plan	Programs	Projects	Aims	Application zone	Stakeholders
Integrated watershed management	Scientific investigation	Inventory of native vegetation	Select vegetation to increase infiltration	Andean piedmont and premountain	University, municipality and sectoral representatives
			Select vegetation to clean the air	Andean piedmont and premountain	University, municipality and sectoral representatives
		Carrying capacity of canals	Storm flow drainage	San Carlos and Las Perdices canals	University, municipality and sectoral representatives
		Piedmont land use capacity	Suitable land use and cover	Andean piedmont	University, municipality and sectoral representatives
		Biodiversity of conservation areas	Protection of valuable ecosystems	Andean piedmont and premountain	University, municipality and sectoral representatives
		Flood protection	Minimise flood hazards	Andean piedmont and premountain	University, municipality and sectoral representatives
	Creek protection	Vegetation conservation	Native vegetation conservation	Andean piedmont and premountain	Municipality, local communities and National Forestry Commission
		Afforestation	Soil and flood protection	Andean piedmont and premountain	Municipality, local communities and National Forestry Commission
	Lowlands protection	Vegetation of riparian buffers near canals	Retain land overflow and increase water infiltration	San Carlos and Las Perdices canals	Municipality and National Forestry Commission
		Increase hydraulic capacity of canals	Increase drainage capacity	San Carlos and Las Perdices canals	Municipality, Public Works Ministry, farmers and canal associations

Table 9.2　Proposed initiatives and projects of the new environmental office (continued)

Integrated watershed management	Community nature experience	Implementation of mountain footpath	Increase social–nature interaction	Andean piedmont and premountain	University, municipality, local communities and NGOs
		Tourism and recreation circuits	Increase local community socio-economic benefits	Andean piedmont and premountain	University, municipality, local communities and NGOs
		Urban sprawl regulation	Urban development planning and management	Peñalolén commune	Municipality and vicinal associations
		Environmental education and training	Increase local knowledge and participation	Peñalolén commune	University, municipality, local communities and NGOs
		Ecological park implementation	Facilitate social–natural experience	Andean piedmont	University, municipality, local communities and NGOs
Institutional strength	Environmental management system implementation	Environmental management municipal department	Design and implement a municipal environmental department	Municipal council and Mayor's Office	Municipal council and Mayor's Office
			Inventory of communal environmental issues	Municipal department and vicinal association	Municipal council, vicinal associations and Mayor's Office
			Definition of communal environmental policy	Municipal department and vicinal association	Municipal council, vicinal associations and NGOs
			Identification of communal environmental goals	Municipal department and vicinal association	Municipal council, vicinal associations and NGOs
			Development of communal environmental programs	Municipal department and vicinal association	Municipal council, vicinal associations and NGOs
			Assessment of environmental performance indicators	Municipal department and vicinal association	Municipal council, vicinal associations and NGOs

Table 9.2   Proposed initiatives and projects of the new environmental office (continued)

Institutional strength	Improvement of communication between municipality and stakeholders	Partnership strength	Increase linkages between municipality and regional authorities	Municipal planning office, waste services, health department	Regional environmental commission and government
			Increase linkages between municipality, professional organisations and university	Municipal planning office	Municipal council, professional organisations and university
			Increase linkages between environmental and local development organisations	Municipal planning office	Municipal council, vicinal associations and NGOs
			Increase public–private partnerships	Municipal planning office	Public services and private organisations
			Increase linkages between municipality and vicinal associations	Municipal planning office	Municipal council, vicinal associations and NGOs
		Updating of environmental municipal regulation	Updating of environmental municipal ordinances	Peñalolén commune	Municipal council and Mayor's Office
			Organisation of round tables and dialogues with local associations	Peñalolén commune	Municipal council, vicinal associations and NGOs
		Awareness and training of municipal professionals	Education on local environmental problem-solving	Municipal department	Municipal department, university and NGOs
		Release and diffusion of environmental concerns	Press campaigns, forums and environmental fairs	Municipal departments	Municipal department, university, vicinal associations and NGOs

millimetres, but when La Niña takes place, the annual rainfall could be just above 100 millimetres (Romero 1985). Rainstorm flooding, large soil erosion and mass earth movements are natural hazards that affect the city each year. These natural hazards should be taken into consideration when preparing any urban suitability plans, especially on the western side of Santiago's watershed where the Pudahuel community is located.

At the western border of the watershed, the coastal range constitutes another source of natural hazards. The main mountain peaks are above 1000 metres, with many short and steep streams that activate with storm flows. Since floodplains have been used mainly for agricultural purposes, irrigation canals are also an important flood source when intense rainfalls occur. They are used to conduct excess discharge.

The study undertaken by first-year geography students considered three main factors for natural hazards for the Pudahuel area: steep slopes (of less than 30 per cent), superficial drainage density, and land use and land cover. In this exercise, suitable urban areas were classified as those areas located in the floodplain and in slopes of less than 30 per cent. These can be seen on the resultant grid (Figure 9.14). Unsuitable urban land is due to the steeper slopes.

In terms of drainage patterns, a flooding buffer of 100 metres was incorporated into the irrigation canal distribution, and a buffer of 200 metres was considered

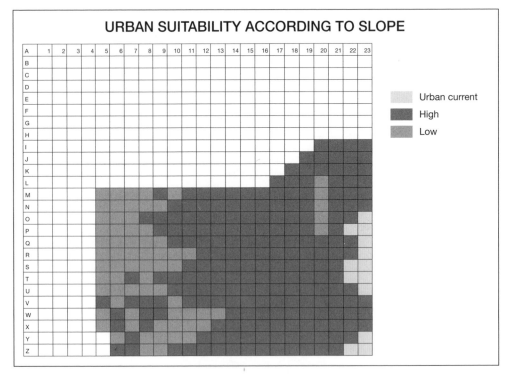

Figure 9.14 Urban suitability according to slope

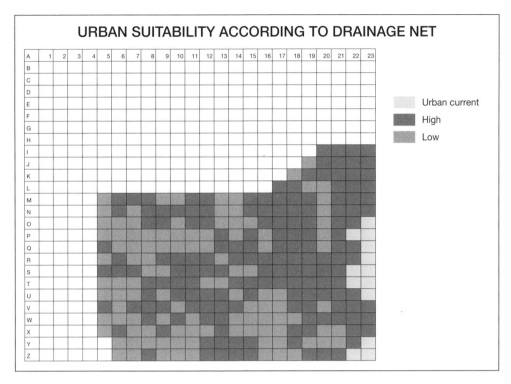

Figure 9.15 Urban suitability according to drainage net

along the borders of rivers and large streams. Figure 9.15 shows the spatial grid of superficial drainage density. Urban-suitable areas have less than three drains per grid square. Unsuitable areas have between four and ten drains per grid square.

Those areas with slopes greater than 30 per cent, and covered by Mediterranean low-density wood forests and shrubs, must be considered unsuitable for urban purposes. Given the poor vegetation cover that is found on steep slopes, there are no urban-suitable areas on the coastal range. Urbanisation should be avoided on the slopes of mountains and hills. Only floodplains and flat areas located beyond the flood buffer protection zones are suitable for urbanisation.

4.2 The impact of the Santiago International Airport on the urban suitability of residential areas

The Santiago International Airport was built in 1967, replacing an older, smaller airport that was historically surrounded by dwellings. When the airport was built all the surrounding areas were devoted to agricultural land use or covered with native vegetation. The city boundary was far enough away to suppose that any direct environmental impact of the airport on the inhabitants and the neighbourhood areas would be prevented by regulations against urban use. Nobody could guess at that time that the city sprawl was going to approach the airport in the way that it has, and in such a short space of time.

However, during the last 40 years the airport has grown parallel to socioeconomic national development. Chile is a fast-growing export-oriented economy which is based mainly on the export of natural resources and raw materials to developed nations. National and international cargo and passenger flights have increased substantially in recent years. New installations, including a second airfield, hotels and airport services, have been developed.

From the point of view of the geography students, it was assumed that all these new constructions had negative impacts, or would impact severely on the local people who live near the airport. According to environmental impact assessments, the main negative impact of the airport on local people is related to increasing noise and air pollution caused by the number and timing of flights. The students analysed some of the environmental issues to find out how the local population has been affected by the expansion of the Santiago International Airport.

The students interviewed 50 families who live as near as possible to the airport. Local perception was sought, assuming that the people who lived there would easily recognise the negative impacts. The aims of the survey were to:

- validate the real meaning and functioning of regulations that have prevented the urban occupation of areas situated very near the airport, and to observe if the present urban boundaries have effectively protected the health of local people
- identify the local people's perception regarding airport environmental impacts
- assess the benefits that the airport has produced on the quality of living of local inhabitants, e.g. in terms of employment and accessibility
- evaluate if noise is the main pollution caused by aircraft.

The main reasons people live near the airport were as follows:

- Availability of jobs was the main reason for 40 per cent, although their sources of employment could be completely independent of airport operation.
- Social housing programs and self-construction located near the airport accounted for 28 per cent of the reasons
- The remainder of the population mentioned tradition as a reason to live in these unsuitable areas. However, the spatial permanence of poor people in the same part of the city for a long period of time seems to be more related to the impossibility of moving to other more suitable areas—which is probably a consequence of lower incomes rather than traditional values.

Most of the people do not perceive any real inconvenience from the airport operation or, if they perceive any issue at all, it is much more related to road traffic such as the increasing number of cars, buses and lorries. Noise and air pollution are mentioned as relevant issues, but again they are related mainly to road traffic, rather than aircraft. According to most of the local people, the airport does not cause any real environmental problems.

The indifference of the people regarding the presence of the airport could be explained by the fact that, for most of them, the airport has not produced any real benefit. Only a modest 9 per cent think that the airport has brought more employment. In terms of public health, 70 per cent of the local inhabitants estimate that the airport has not caused any specific illness and, as a final conclusion, 70 per cent of the people do not suggest the airport should be moved.

This exercise allowed the students to learn about the differences between scientific geographical reasoning and common sense. Geographers formulate hypotheses and then go into the field to test these assumptions and to find out the perception of local people and how they view the environmental impact of large development projects, such as international airports.

The students were initially disappointed with the result of this exercise because it opposed their initial judgments, but they later realised the necessity of fieldwork in providing as much information as possible about local people and cultures.

4.3 Selecting sites for nature conservation and recreation in Pudahuel commune

Recreational and nature conservation areas are urgently needed in Santiago city, and especially in the Pudahuel area (CONAMA RM 2002). Given the high concentration of pollution (air, water and soil), the number of illegal waste deposits and the lack of green spaces in the urban area, this commune has to integrate its rural areas with its sustainable development. Those rural areas are the slopes, valleys and foothills of the high coastal range.

Coastal range areas should be integrated with communal development. At the present they remain marginal and unproductive lands that are located beyond the urban boundaries. However, coastal range slopes and foothills have outstanding ecological areas and pleasing rural landscapes. One way to integrate these areas with local development is by creating nature conservation and recreational zones.

In this exercise, first-year students looked for some areas in the Santiago coastal range that could be allocated to conservation and recreational purposes. To select suitable sites, they took the following steps:

1   First, they considered the road classification system, mapping the existence of highways, paved and unpaved roads, and urban and rural streets and footpaths. They were particularly interested in the accessibility of proposed areas, using good-quality roads to ensure visitors from nearby urban areas (Caviares 1999).
2   Second, they considered a map of contour lines, recognising that steep slopes could limit most of the accessible recreational and conservation areas. Steep

slopes are a severe limitation for accessibility, and increase the risk of soil erosion and natural hazards.

3   Finally, they considered land use and land cover, because the selected areas should have natural vegetation and adequate cover to offer enough shadow, beautiful landscapes, protection of biodiversity, and protection against soil erosion and water pollution.

By spatially integrating these geographical variables, they produced a map showing the selection of the most suitable areas, according to the chosen variables. These areas combine easily accessible sites, located mainly on the coastal range piedmonts, and in foothills that are well-protected by native forests and shrubs.

During fieldwork, students talked with local people and interviewed local leaders. They analysed the limitations of a geographical approach purely based on physical aspects. They realised that all local people want to be involved in economic planning to develop recreation and ecotourism. They need to combine modest incomes so that they can move from a traditional rural life to one with complementary activities to take advantage of having a large city nearby. However, to achieve this purpose, they need finance, training and social support, and a new local development program, where their traditional way of living can be integrated to modern economic activities. Unfortunately, they do not have the institutions to strengthen their specific knowledge and skills.

5. Conclusion

The exercise strengthened the ecological meaning of geography for the students, and stimulated their awareness of ecological values. After the study, many of them wanted to participate in voluntary ecological movements, and generate non-government organisations (NGOs) much more involved in nature conservation. They also realised the need to integrate physical and human geography, and to relate global and local scales, to offer sustainable development plans. Finally, they realised the important role that they can play in political geography, suggesting and proposing new institutions and organisations to cope with sustainable development challenges.

Acknowledgment: The figures, tables and photographs in this chapter were prepared by students in the Environmental Planning and Management Magister Program, University of Chile. Reproduced with permission.

Bibliography

Aravena, D 1994, *Determinación de áreas de riesgo en la comuna de Peñalolén*, 1994, Seminario Facultad de Geografía, Universidad Católica de Chile.

Arnold, C & Gibbons, J 1996, 'Impervious surface coverage: The emergence of a key environmental indicator', *Journal of the American Planning Association*, vol. 62, no. 2, pp. 243–258.

Barnes, K, Morgan, J & Roberge, M 2001, *Impervious surfaces and the quality of natural and built environments*, Department of Geography and Environmental Planning, Towson University, Baltimore.

Caviares, A 1999, 'Calidad de vida, recreacion y biodiversidad', *Revista Ambiente y Desarrollo*, vol. XVIII, pp. 44–50.

CONAF–CONAMA 1999, *Catastro y evaluación de recursos vegetacionales nativos de Chile*, Chile.

CONAMA RM 2002, *Areas verdes en el Gran Santiago. Area de ordenamiento territorial y recursos naturales*, Impresiones Nordicas, Santiago de Chile.

Corvalán, P, Kovacic, I & Muñoz, O 1997, 'Quebrada Macul: el aluvión del 3 de mayo de 1993, causas y consecuencias', *Revista Geográfica de Chile Terra Australis*, no. 42.

Espinoza, G, Valenzuela, F, Jure, J, Toledo, F, Praus, S & Pisan, P 2000, *Gestión municipal: políticas, planes y programas ambientales: Experiencias en los municipios de Alhué, El Bosque y Lampa*, Centro de Estudios para el Desarrollo (CED), Ediciones del segundo centenario, Santiago, Chile.

I. Municipalidad de Peñalolén 2001, Plan de Desarrollo Comunal (PLADECO) 2001–2005, de la comuna de Peñalolén.

Instituto Nacional de Estadística (INE) 2002, Censo de Población y de Vivienda del año 2002, INE, Madrid.

Romero, H 1985, *Geografía de los climas de Chile: Tomo XI coleccion de geografia de Chile*, Editorial Instituto Geografico Militar (IGM).

Romero, H 2004, 'Crecimiento Espacial de Santiago entre 1989 y 2003 y la Pérdida de Servicios Ambientales', in Patricio Tupper (ed.), *Hacer Ciudad*, Centro Chileno de Urbanismo, Santiago, pp. 179-201.

Romero, H, Ihl, M, Rivera, A, Zalazar, P & Azocar, P 1999, 'Rapid urban growth, land use and air pollution in Santiago de Chile', *Proceedings from International Conference on Urban Climatology*, Essen, Germany, Atmospheric Environment, vol. 33, nos. 24–25, pp. 4039–4048.

Romero, H, Vásquez, A & Ordenes, F 2003, 'Ordenamiento Territorial y Desarrollo Sustentable a Escala Regional, Ciudad de Santiago y Ciudades Intermedias en Chile', in Figueroa, E & Simonetti, J, 2003, *Globalización y Biodiversidad: Oportunidades y Desafíos Para la Sociedad Chilena*, Vicerrectoría de Investigación y Desarrollo de la Universidad de Chile, Santiago, Chile: Editorial Universitaria, pp. 167–224.

Romero, H & Vásquez, A 2005a, 'La Comodificación de los espacios urbanizables y la degradación ambiental en Chile', *Scripta Nova*, Revista Electrónica de Geografía y Ciencias Sociales, Universidad de Barcelona, vol. IX, no. 194, 1 agosto 2005.

Romero, H & Vásquez, A 2005b, 'Evaluación Ambiental de las Cuencas Urbanas del Piedemonte Andino de Santiago de Chile', *Revista EURE de Estudios Urbanos*

Regionales, Pontificia Universidad Católica de Chile, vol. XXXI, no. 94, Diciembre 2005, pp. 97–118.

Zandbergen, P, Schreier, H, Brown, S, Hall, K & Bestbier, R 2000, *Urban watershed management* version 2.0, CD-ROM, Institute for Resources and Environment, University of British Columbia, Vancouver.

Notes

1   This chapter has been supported by an IGU Grant and the Chilean National Scientific Fund Nº1050423.

Community and sustainable development: Sustainability awareness through transportation, food and water in a Buenos Aires neighbourhood

10

Gabriel Fabián Bautista

1. Introduction

The Buenos Aires sustainability project's implementation was partially developed from September to December 2005. It grew out of participation in a three-week professional learning course held in Rome during June 2005. This chapter describes three sustainability subprojects that were undertaken, covering transportation, food and water. These three aspects form part of a system, and include the full range of the population. The food subproject involved children and their parents; the transportation subproject involved the adults, or economically active population; and the water subproject involved the elderly. These three subprojects were carried out in Villa Urquiza, which is a neighbourhood within the Buenos Aires metropolitan area, home to 14 million inhabitants.

In Figure 10.1 (colour section, opp. p. 224), Buenos Aires is seen as a light grey-brown patch on the right side of the Río de la Plata. The tentacles of water entering into the pampas are clearly seen, as is urban expansion. However, as discussed below, this grey area does not reflect the whole truth, because there are suburban neighbourhoods that are no longer rural and yet the image shows them as such. Urban sprawl is significant to the understanding of transportation, food and water issues in this region.

The pampas is a very flat and fertile plain with drainage problems even in rural areas.

Villa Urquiza is 7 kilometres from the river, which is the main source of water and the place where the city watersheds collect. Most of these, except the Riachuelo in the south and the Reconquista in the north, have been rebuilt through drainage pipes.

Villa Urquiza *barrio*, or neighbourhood, is located in the city of Buenos Aires, capital of Argentina. It lies between the barrios of Villa Pueyrredón, Belgrano,

Ortúzar, Coghlan, Saavedra, Parque Chas and Agronomía. Its limits are the streets and avenues named Constituyentes, Crisólogo Larralde, Galván, Núñez, Tronador, Roosevelt, Rómulo S. Naón and La Pampa.

This is a neighbourhood still full of old houses, but due to real estate price rises they are being demolished to build apartment buildings, making the population more dense in the area. This ongoing process makes heavy demands on transportation, food and water. The streets are still relatively quiet, but the avenues are crowded with fast traffic. The area has some parks on its borders, and some squares, but they are neither close enough nor big enough for the population. However, it still is quite pleasant compared with some other neighbourhoods, and during the summer one can see many neighbours talking to each other, and comfortably sitting on chairs on the sidewalk. Insecurity, impersonal buildings and changing habits in the newcomers are now starting to bring about change.

The three subprojects have a geographical setting in an urban community. The processes of change have led to environmentally harmful transportation, and unsustainable food habits. The urban–rural interface, which is expanding, has also caused flooding, a reminder of the natural–urban landscape forces. The three projects have involved the parish community of Nuestra Señora de la Anunciación in the Villa Urquiza neighbourhood, and contacts with municipal authorities and non-government organisations. The pedagogic tools are based on personal interaction, such as group meetings, storytelling, mapping, fieldwork, and an information session using brochures, poster presentation and informal discussion.

2. Subproject 1: Transportation issues in a Buenos Aires neighbourhood—how 'far' is 'close'?

Transportation issues are a key factor for sustainability in urban areas and they are linked to the use of natural resources such as oil from the country itself, from neighbouring countries and from foreign countries as well. Transportation is a key factor in neighbourhood sustainability, for it can maintain healthy relationships within a community. Transportation allows people to live in one neighbourhood yet still work in another, and it helps to support the livelihood of the people living there with small shops and services, including health services.

The need to reach the outer limits of the metropolitan areas inevitably perforates the web of life of existing neighbourhoods, and the interventions change lifestyles, property, commercial affairs and culture. There is also the issue of choice in the way we travel, how we develop citizenship in terms of controlling what the municipal authorities decide to do, and how citizens in the neighbourhood observe their own changing habits.

The expected outcome was to develop conscious choices in the way we travel, to initiate citizen participation, and to change habits. The change in habits is crucial for energy saving, especially if we think in terms of carbon dioxide and greenhouse gases that affect our lives on earth. The learning experiences included mapping and fieldwork. The target group was the economically active population, as they are most compelled to go out to work and need mobility. This mobility is a particular feature of human population related to technological issues. Therefore, people reflected on the way they travel and the effects of the choices they made.

Today, children and the elderly are becoming more and more mobile in their spare time. The project members made a comparative field study, travelling by different means of transportation, mapping the itinerary and reflecting upon the environmental issues related to the choices made. The methodology and evaluation was conducted through group meetings and interviews, and by sharing within the group and with the community their personal stories of daily travel. The experience was very enriching for all of us. There was a brochure produced and artwork was exhibited in the discussion session. The key questions were:

- Where do we live?
- What is the transportation like?
- What are the times and the choices available?
- Why does it take so long to travel from home to work?
- How does our travelling influence the urban environment, and what can we do about it?

2.1 Transportation subproject aims

Transportation could be the major sustainable issue in a metropolitan area and the aim of the subproject is to make people aware of the effects of transportation choices. Buenos Aires is the capital of Argentina, and the country suffers from what has been called macrocephalia (since more than half of the country's population is living in the metropolitan area), where every action concerning transportation is somehow related to a national scale. This is true in the case of the Buenos Aires waterfront area where investment is maximised by trading ports, the national airport, the highways and avenues, important corporate buildings and the recycled old city port area which has become a new frontier of land within the city. The process of gentrification is also a factor that is gaining momentum.

This context affects the way people are able to travel from the neighbourhoods to work. It also affects the choices they make in everyday transportation. There is a link between urban sprawl and the unsustainable use of private transportation. Urban sprawl has affected water absorption and the landscape, leading to more frequent flooding, and affecting the production and consumption of food. It all relates to the economic system that encourages the production and use of private transportation; more roads; changing land use patterns; and urban, peri-urban, suburban and rural real estate business.

At the regional level, the aim is to change transportation habits from private use to public use. The Buenos Aires metropolitan region houses 14 million people, so transportation is a key factor related to a sustainable future in the city. Huge investments have been made in private transportation, but less for public transportation, especially trains and buses. The subway has finally reached Villa Urquiza, which is going to be a transfer point to go downtown from the suburbs. It will open to the public in 2008.

The community is ambivalent about the subway. The commercial sector feels favoured, yet is worried about parking problems. The residential sector feels worried about more noise, pollution, robberies and insecurity, yet residents are at the same time happy because their property values have been rising and they will also be able to reach the downtown area in 20 minutes.

At the local level, the group was conscious about the type of choices made and the impact on energy saving, especially in terms of carbon dioxide and other greenhouse gases. The economically active population goes to work and wants to travel faster, more easily and more comfortably. The subway, train or bus might not be the most comfortable way of travelling, but they are faster and more sustainable. Workers use their cars to travel comfortably but transfer at some point on their way to work. They seldom go all the way downtown with a car.

The aims of the project were accomplished with a sample group of the parish community. They were able to map their daily travelling itineraries and the different alternatives. They produced a poster session for the rest of the community to ponder about the connections between travel choices and the environment. They told stories of travel to and from work everyday. Artwork related to this helped the community ponder travel methods and the personal connections with humans. The reflections of Yi-Fu Tuan provided guidance (see Tuan 2007).

2.2 Knowledge outcomes

The parish group undertook a comparative field study about travelling by different means of transportation, mapping itineraries and reflecting upon the environmental issues related to choice, such as energy saving, environmental impact, noise pollution, human relationships, and the links between the urban ecosystem and the natural ecosystem. They also learned to interpret satellite images and acquired a greater consciousness of the sense of community.

This group mostly comprised the economically active population, and younger and older relatives. Fieldwork, mapping, group meetings and listening to the stories people told about their daily travels to and from work was important because it heightened awareness of this daily practice and its impact at regional and national levels. Interviews with key informants such as clerks and young people employed in different services, and the way they told their stories about travelling within the city, were relevant to this project.

2.3 The local social context

The learners or target group were the general public, particularly the economically active population between 18 and 65 years of age. However, children and the elderly have become more mobile due to the changing uses of spare time. The school population, non-government organisations, local municipal government bodies and university students were consulted. The direct political context targets local authorities and transportation entrepreneurs. There is plenty of information disseminated by research centres and the bureaucracy, but it is hard to access because transportation is such a major business enterprise.

The parish group developed maps, reflected about the way they travel, compared time and money, made a poster and presented it. They learned skills on how to gather information and interpret it, to map, and to use computer technology to make a presentation. The weakness of the project was that it was probably too locally based. However, in a second phase, the group is hoping to transfer the experience to other parish communities of Buenos Aires.

2.4 The learning experiences

The learning experiences and activities included group meetings, storytelling, mapping, fieldwork (including gathering information from newspapers) and a poster presentation. The participants reported that they had a more developed awareness of the place, of the sense of place, and the geographical significance of distances and position. The evidence upon which judgments were based was from fieldwork, mapping and the quality of the communication of the experience.

This learning experience is transferable to other urban areas, especially metropolitan areas or medium-sized cities. There is a need for mobility when going to work and this need is expressed by access to transportation, so it is possible to reflect on it, map it, and consider the consequences that transportation has on the environment. It contributes to the understanding of sustainability because transportation choice is a clear component of the environmental system, related to the emission of greenhouse gases and global warming. The language and culture is irrelevant because transportation is universal. The project gives a sense of solidarity, since everyday travelling is crucial in every neighbourhood, especially in urban areas, and it also expresses a willingness to make more sustainable choices in regard to others and the problems of the earth.

Finally, one may say that these subproject outcomes were successful because a good brochure was produced, relevant and effective fieldwork was undertaken, a clearer consciousness on how transportation choices are connected to the environment was developed, and a willingness to change to more sustainable ways of travelling was demonstrated.

3. Subproject 2: Food consumption in a neighbourhood of Buenos Aires—is it true that we are what we eat?

This project is also located in the Villa Urquiza neighbourhood of Buenos Aires. It expresses the conflicting society–nature relationship, and the interaction of rural and urban food habits. The sixth-grade group in the study became more aware of the connection between their food consumption and the green belts and transportation, which is in turn related to urban sprawl and globalisation. (Urban sprawl also provides a direct link to the next subproject about water, changing landscape, and flooding.)

Food habits remain the principal link with the land and ocean, as an integral part of our home on earth. Grocery stores and supermarkets could nowadays be considered as the main link with the land. It is in this sense that the choice we make when we eat simple foods for our daily lunch or supper is not an occasional choice, but one that has its own weight because it is related to supporting local food producers instead of favouring the agro-industry capital concentration. It is in this sense that the rural and urban meet whenever we eat.

These conscious choices in the way we eat, what we eat and what we buy are an integral part of the rural and the urban mix. Not only does our survival depend on these choices, but they are also an expression of our relationship with the earth in general and with our land in particular. They also convey a significant cultural sense in terms of the meaning of having a meal.

The learning experiences were especially related to sixth-grade children and their parents, because parents usually are the ones who buy the food at the grocery store or the supermarket. Part of the learning experience related to sustainability and the impact our choice of foods has on it. The children were able to develop a clear understanding of what sustainability means, especially in regard to the rural sector and its capacity to keep producing enough food now and for future generations. They pondered their weekly menu and made a comparison between the older and younger generations, and the consequences that eating habits have on the relationship between rural and urban living.

The children became aware of the food security topic through the knowledge that green belts are disappearing because of urban sprawl. They also learned about the need for a bigger rural agro-industry, a bigger scale of production for future needs, and the dangers of monoculture (for example, soy beans). How these affected the quality of soil, water and air and could lead to more frequent flooding and the degradation of neighbourhoods was studied. They were able to develop a system-thinking process as a game.

3.1 Food subproject aims

The principal aim was to develop in children an awareness of the connection between what they eat and the chance for a sustainable future. In the local context it means supporting the small grocery store as a way of avoiding neighbourhood decay. The aim was only partially accomplished, because children do not buy the supplies by themselves; their parents do, but they had conversations with them about it.

In a national context, the rural sector plays a major role, since its economic activity is the most important economic activity of the country, and the one that contributes most to gross domestic product. It is also important in terms of population, because rural areas are where 40 per cent of the Argentinean population lives. The aim here is a radical contribution to the national environment. In a regional and local context, the children in Villa Urquiza have become more aware of the connection between the rural and the urban. Knowledge and experience of the fringe or the peri-urban area has a decisive role to play in terms of food production and sustainability. Since the children travel throughout the urban and the suburban area, and some have a second home or a cottage, their memories and vacation trips have proved significant when thinking about these issues.

The main project aim was to develop conscious choices about what children eat. It is important that they know that food is for sustenance, and is also an expression of the relationship with the earth in general and with their land in particular.

3.2 Knowledge outcomes

The San Alfonso Catholic School sixth-grade children produced the draft of a meal plan, reflected on it, and shared this with their parents and the other grades. This storytelling process has contributed to sustainable education. They have also understood what type of advertisement promotes what type of food. They have acquired a deeper knowledge about the relationship between food, the system to produce it and sustainability by mapping and graphing the links that facilitate food from its purchase at the store to the consumption of the meal.

The children were able to develop a system-thinking approach as opposed to fragmentary knowledge. They thought of themselves as part of the urban population, and yet very much related to the rural, as well as to transportation and water issues. A second phase will be to inform local authorities and the local community with a poster session and present a written statement about an issue concerning food consumption.

3.3 The local social context

The local social context is the sixth-grade low- to middle-class children of a Catholic school. The target groups in this case are these children and their parents. By studying their way of eating (by looking at the components of a meal), they considered whether they are eating in a way that favours sustainable development for the future, or are producing a tendency towards unsustainability.

The children were able to formulate how a study of food broadens into the more general question on how rural life is evolving in Argentina and the people involved in this process of change, a process that means capital concentration, and a pattern of land tenure that excludes the traditional small producers and favours the bigger ones. They were able to understand about the marketing of food, about supermarkets and grocery stores, how food is offered to the public and consumer, and how lifestyles are related to cooking and eating. Finally, the children were able to incorporate family issues, because this process involves not only knowledge, but values. Some of them said that they had to cook for themselves, their brothers or sisters because they were alone at lunchtime, and sometimes even at supper because either their parents were divorced and their parent had to go out to work, or both parents had to go to work to earn enough money to pay the bills. In some cases they had 'fast food' delivered, even though their parents were at home.

An interview was also conduced with the chair of a *comedor infantil*, sponsored by the parish Jesús Misericordioso, which is a place where poor children eat lunch from Monday through Friday. These children belong to the same neighbourhood of Villa Urquiza, but they live in a housing area that has been reclaimed. The houses were going to be demolished to build a highway, but this was not built. Since they were empty, immigrants from rural areas and from neighbouring countries occupied them over a period of 20 years. Paradoxically, these children ate a more diverse menu than others, and yet they showed the same preferences—they all liked junk food, but were not used to eating vegetables or boiled potatoes.

In a second phase there ought to be a closer contact with the municipal government, the local grocery owners and the huge supermarkets, and non-government organisations like the Grupo de Reflexión Rural, which has developed a political approach to food production and consumption, and its relationship with the soy-bean monoculture as a commodity.

3.4 The learning experiences

Key questions asked were:

- What do we eat?
- Where does the food come from?
- Why we are eating what we eat?
- Is it sustainable or not?
- What does our weekly menu comprise?
- How could it be different?

To answer these questions, the children thought about them at home and in the classroom. Apart from maps, graphs and posters, they learnt how to develop interviews, conduct group sessions, distribute material, explain their aims to authorities, and publicise the experience. The meetings were held in the classroom. The children developed the capacity to formulate the results of discussions, comments, notes and pictures. Yet the main skill learned was political involvement. The children got different perspectives on these issues, increased their participation, and aspired to a better quality of life.

The weakness of this subproject was similar to the first—the limited size of the learning community. Yet this was also a strength since it developed much greater involvement. In a second phase, the wider group would include more neighbourhood people supported by professionals, and it should also include the local political context (such as more participation by the people), developing citizenship, and the involvement and commitment of local authorities, school authorities and church authorities (particularly the parish community and the parishes of Buenos Aires).

The final objective of achieving a profound change in the eating habits of children was partially achieved. Although children's eating habits were not profoundly altered, their awareness of this issue has been raised. Children also understand the city–country relationship; that is the production, distribution and final consumption of food. Daniel Kemmis' *Community and the politics of place* (1990), which emphasises the importance of active and involved citizenship in environmental issues and the aspects of belonging to a place, helped to sensitise local authorities and the local community to the issues. They learned the hardships of citizenship through a poster session, publishing a menu and advertisements promoting various kinds of food.

The judgments made by the children were based on interviews, group meetings and discussion on a weekly topic. The project is easy to transfer to other regional or national contexts since it is quite easy to consider what we eat and where the food comes from, and all the links that bring food to our homes. In some cases, there would be a light mediation process, in others a denser web of mediation processes. The project contributes to the understanding of sustainability by making us aware of the links in food production and how these links impact on the environment, in particular that blurred frontier between the rural and the urban. Food is a multi-cultural language and the reflection on its condition can be shared and understood by any means.

However, the evidence shows that the project outcomes were only partly successful. The children pondered about where food comes from, did good fieldwork and got a clear idea about how food is related to the environment, but they were unable to make much change in how food is purchased.

4. Subproject 3: Flooding in a Buenos Aires neighbourhood—where did nature go?

One of the most dangerous consequences of urban sprawl is flooding in the suburbs of Buenos Aires. Some people are not prepared for this because their perception of nature is incorrect. The urban landscape and the urban setting are usually thought of as a cultural landscape with no further connection to nature, and especially to the hydrological cycle. One could ask whether there is any natural landscape out there at all. There is also a prevailing perception that the Buenos Aires metropolitan area has been built on flat land, but this landscape is not as flat as it may seem.

The successful development of watershed awareness and a more systematic and holistic view of the hydrological system underneath will allow development of new environmental practices, and a better control of the decisions that the municipal government might take in regard to these issues. The awareness of the geomorphology of the urban landscape and its natural systems is also a way of reconnecting people with nature, even in a highly urbanised setting. This issue is crucial to understanding what sustainability means and helps strengthen not only new ways of seeing the conflicting urban–natural relationship, but also developing new practices, such as a more conscious use of water.

The older generation is very important because its members can tell stories about the previous neighbourhood landscape. Telling these stories to the younger generation is a crucial point in developing an intimacy with the landscape as it exists now. This narration of the story must be complemented with fieldwork, because it is through this kind of fieldwork that we learn about the landscape now, as it was decades ago, and even as it was centuries ago, since the city was first established by Spanish settlers 450 years ago. This connection between the older and younger generations is a crucial learning experience.

This project was only partially completed at time of writing.

4.1 Water subproject aims

The objective is to develop a clear vision of how the hydrological cycle functions in the neighbourhood and its relationship to flooding and broader environmental connections, such as greenhouse gases. This connects it to subproject 1 (transportation) and subproject 2 (the rural interface of food, green belts and the peri-urban frontier). Some interviews were conducted, and the main questions were:

- Where are the watersheds?
- How does the rain flow on a rainy day?
- Why is there flooding?
- In what condition is the landscape system?
- Why and how has the landscape become this way?
- Why is it important for our health and for the quality of drinking water?

The broad aim was to consider water issues and flooding in other parts of the country. At the regional level, the aim was to learn about watershed consciousness in the metropolitan region. At the local level, the community was asked to reflect upon their memories about the watershed and transfer this to the younger generation. The bioregional experience might help to renew watershed awareness.

4.2 Knowledge outcomes

The main outcome was to tell the story of how the landscape and the region have evolved and the problems that this evolution might cause to the hydrological condition of the neighbourhood. Maps, drawings or old pictures of the landscape, in particular the hydrological conditions as they were in the past, should be used to provide comparisons between past and present conditions.

The local community and local government are expected to engage in dialogue. The local community will have better tools to understand the place where they are living and develop a more sustainable way of life. Making sense of water issues, not only in terms of domestic consumption but also in how the whole hydrological network works within such a strongly built environment, is the aim.

4.3 The local social context

The older generation remembers the neighbourhood landscape as it was. By telling their children and grandchildren how the landscape used to be, the younger generations learn how the landscape has changed and the role of water within it.

4.4 The learning experiences

These include:

- mapping and graphing the landscape
- comparing the past with the present
- telling the story of landscape change
- organising knowledge in a systematic way
- communicating the experience
- formulating clearly what is required to keep the landscape safe and healthy for the future.

The subproject evaluation is based on:

- how the younger generation is able to tell the story that their elders have told them about the previous landscape
- the mapping of environmental history (with especial emphasis on hydrology)
- systems assessments
- the production of graphics that show the connections between drinking water and the urban environment.

The project is transferable to other local contexts, regions and countries. It can develop a deep awareness of the hydrological cycle and natural hydrology and contribute to the understanding of sustainability. Water is a key issue in sustainability.

It is one of the most important community issues. We rely even on water being there when we need it.

The project's outcomes are successful if there is a better knowledge of landscape, an increased intimacy and sense of the place, a good brochure design, thorough fieldwork, and a clear consciousness on how water-related choices and behaviours are connected to the environment.

5. Concluding remarks

The project, composed of three subprojects, involved a system-thinking process that was partially accomplished. In the third phase of the project, the community will gather together and make a clearer connection between transportation, food and water as an important force for changing and modelling the landscape.

Bibliography

Kemmis, D 1990, *Community and the politics of place*, University of Oklahoma Press, Norman, OK.

NASA Johnson Space Center – Earth Sciences and Image Analysis (NASA-JSC-ES&IA) 2003, *Uruguay/Rio de la Plata, Buenos Aires*, satellite image, 17 March, National Aeronautics and Space Administration, viewed 27 June 2007, (http://nix.ksc.nasa. gov/info;jsessionid=7bq790q9gefqt?id=ISS006-E-38952&orgid=3).

Tuan, Y 1974, *Topophilia: a study of environmental perception, attitudes, and values*, Prentice Hall, Englewood Cliffs, N.J.

Tuan, Y 2007, *Yi-Fu Tuan*, viewed 27 June 2007, (http://www.geography.wisc.edu/~yifutuan/).

11 | The forest fires issue in Portugal

Manuela Malheiro Ferreira and Jorge Duarte

1. Introduction

In Portugal the forest area occupies 5.4 million hectares, which is about two-thirds of the area of the continental part of the country. Between 1981 and 2003 the Portuguese forest area increased by about 460 000 hectares (an average of 20 000 hectares per year). This was a result of the reforestation of agricultural lands, supported by the European Community rules that aim to increase a country's forest cover. The mix of species is represented, in decreasing order, by *Pinus pinaster* Ait., *Eucalyptus* spp., *Pinus pinea*, *Quercus suber*, and with a lesser representation by other coniferous and broadleaf species.

The issue of forest fires in Portugal is a very important one. In the 20 years between 1985 and 2005 there were 12 years where the burnt area exceeded 100 000 hectares. In 2003 the burnt area reached a massive 425 716 hectares and in 2005 it was 293 911 hectares.

In 2003, the country faced a heatwave with temperatures over 40°C, very low humidity, intense winds and dry thunderstorms. The meteorological conditions and the smoke from fires generated similar conditions to the greenhouse effect. The superheating kept the atmosphere hot and dry, so no rain fell to extinguish fires. The Resolution of the Council of Ministers 106-B/2003 declared a public disaster and approved special measures and aid. Fires were verified from 20 July 2003 in the districts of Bragança, Guarda, Castelo Branco, Coimbra, Santarém, Portalegre, Leiria and Setúbal. The government also called for civil protection from the European Union, aiming at the immediate cooperation of all state members.

The forest fire risk is influenced by diverse factors that vary according to time and space.

- Portugal has a Mediterranean climate, with a very warm and dry summer.
- Climate change seems to contribute to warmer temperatures.

- Great areas (about 87 per cent) of Portuguese forest are held privately. Many owners have left rural areas and now live in cities, and do not care about the forest, so consequently the undergrowth has not been cleared as it was in the past.
- People cause fires.
- Vigilance in the forest is inadequate.
- There are few roads in forest areas to support firefighting.

The risks was formerly only calculated at local levels. The fires in the summer of 2003 killed eighteen people and destroyed 425 716 hectares of forest, agricultural land and bushland; previously fires averaged a size of about 120 000 hectares per year. The Portuguese Government announced a vast set of measures with the intention of hindering this kind of calamity in the future. On 21 April 2004 the Agency for the Prevention of Forest Fires[1] was created, with responsibilities which had belonged to the General Department of Forest Resources (Direcção Geral dos Recursos Florestais— Divisão de Defesa da Floresta contra Incêndios). In 2004 the burnt area was 129 539 hectares but in 2005 it increased again to 293 911 hectares. This problem of huge fires hinders sustainable development of large areas of Portugal.

A project about forest fires was developed in the middle school of Ceira (Escola Básica 2,3 de Ceira). Ceira is a *freguesia* located in the central area of Portugal in the District and County of Coimbra (Distrito e Concelho de Coimbra).[2]

2. Characterisation of the forest fires in Portugal

Fire is a natural phenomenon in Mediterranean countries and is a *sine qua non* for the natural regeneration of certain plants. Some plants have adapted to fires and their regeneration is quicker after an area is burnt.

According to the Köppen-Geiger climate classification, the largest part of Portugal, including the area of Coimbra, has the climate 'Csa: Humid Subtropical (Mediterranean)' where 'C' indicates a temperate climate with an average temperature of the coldest month under 18°C and above –3°C; 's' indicates that the dry season occurs in the summer of the same hemisphere; and 'a' indicates that the warmest month has an average temperature over 22°C.

Climatic characteristics of Mediterranean areas favour the incidence of forest fires, with high temperatures in summer, low levels of precipitation and high evaporation. This makes vegetation easily flammable due to the dry summer weather.

According to Rebelo (2003), the spontaneous ignition of fire, though possible, rarely occurs. Ignition after a dry thunderstorm can take place, but frequently fire is caused by humans, whether accidental or by arson. In Portugal, this natural phenomenon has become a calamity. In order to show quantitative evidence of the phenomenon, Figure 11.1 illustrates the number of large and small fires (with a burnt area less than 1 hectare) which occurred between 1994 and 2005.

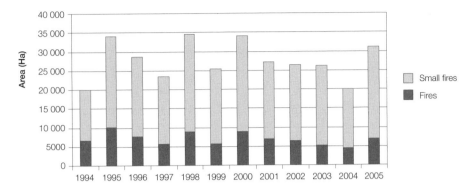

Figure 11.1   Evolution of the forest fires and small forest fires (< 1 hectare) in the last decade

Source: DGRF, DSDF (2005)

The burnt area in the same period (1994–2005) is shown in Figure 11.2.

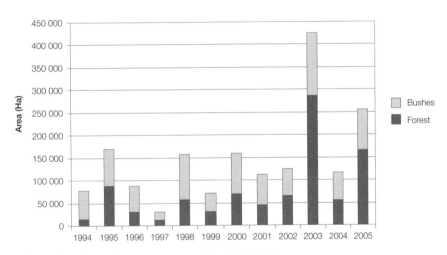

Figure 11.2   Evolution of the burnt area in the last decade

Source: DGRF, DSDF (2005)

There is an oscillation in the size of the burnt area, with a maximum in 2003.

3. Meteorological conditions and the ignition of forest fires

To show evidence of the relation between meteorological conditions and the ignition of forest fires, the year 2005 was chosen because during this year there were winds that affected the entire Portuguese territory, and the burnt area was the second largest in the last decade (see Figure 11.3).

The Portuguese county that registered the largest burnt area was Coimbra—the county where the school is located where this project was developed—so its meteorological conditions were analysed. The data were collected by the Geophysical Institute from the University of Coimbra (Instituto Geofísico da Universidade de Coimbra), and by the National Fire Brigade and Civil Protection Service, Delegation of Coimbra (Serviço Nacional de Bombeiros e Protecção Civil, Delegação de Coimbra). During the analysed period, 202 fires were registered from 1 January to 15 September.

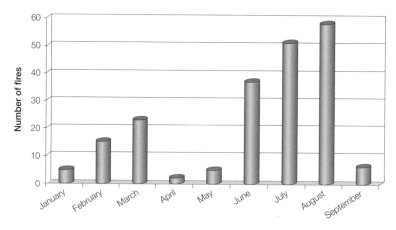

Figure 11.3 Number of fires per month in the county of Coimbra (Concelho de Coimbra) from 1 January to 15 September 2005

Source: National Fire Brigade and Civil Protection Service (2005)

The greatest number of fires occurred in summer time in the months of June, July and August. Nevertheless, in February and March there was an above-average number of fires as a result of severe winds throughout Portugal (the levels of precipitation were very low throughout the year). The meteorological elements analysed were those that seem to have a greater influence on fire ignition and spread, namely temperature and relative humidity.

The frequency of fire ignition by the hours of the day was analysed. The time of the ignition of forest fires in the county of Coimbra is similar to the rest of the country. There are more forest fires during the period of the day with the highest temperatures—between 2 p.m. and 4 p.m. But even during the night, between 9 p.m. and 4 a.m., a small number of fires also occurred.

The frequency of fires by weekday was also analysed. There are no very significant differences but the greatest number of fires occurred from Thursday to Sunday.

A correlative analysis was done between the days with the highest maximum temperatures and the number of fires occurring on these days. Figure 11.4 represents the number of days with a maximal temperature and the number of fire spots occurring during these days. Figure 11.5 represents the correlation between the number of fires occurring and the maximum temperature of the day.

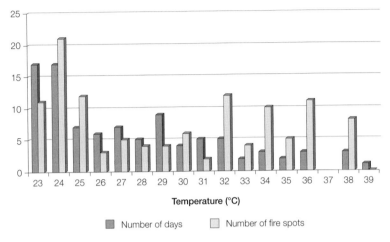

Figure 11.4   Number of days with a maximal temperature and number of fire spots occurring during these days in the county of Coimbra (Concelho de Coimbra) from 1 January to 15 September 2005

Source: Geophysical Institute of the University of Coimbra and National Fire Brigade and Civil Protection Service (2005)

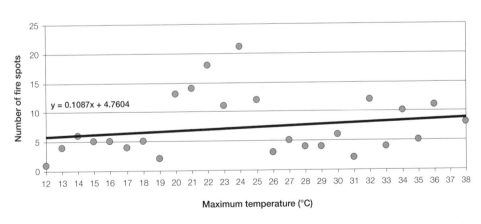

Figure 11.5   Correlation between the number of fire spots and the maximal temperatures of the days they occurred in the county of Coimbra (Concelho de Coimbra) from 1 January to 15 September 2005

Source: Geophysical Institute of the University of Coimbra and National Fire Brigade and Civil Protection Service (2005)

Figures 11.4 and 11.5 show that there is a positive correlation between the maximal temperature and the number of fire spots registered.

Another weather feature favourable to fires occurring in summer is low relative humidity. Figure 11.6 shows the number of fire spots by the relative humidity of the day they occurred.

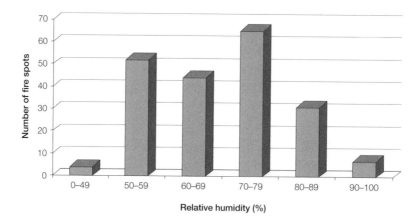

Figure 11.6   Number of fire spots by the irrespective value of minimal relative humidity on the day they occurred in the county of Coimbra (Concelho de Coimbra) from 1 January to 15 September 2005

Source: Geophysical Institute of the University of Coimbra and National Fire Brigade and Civil Protection Service (2005)

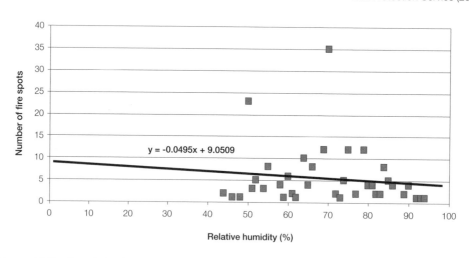

Figure 11.7   Correlation between the minimal value of relative humidity and the number of detected fire spots in the county of Coimbra (Concelho de Coimbra) from 1 January to 15 September 2005

Source: Geophysical Institute of the University of Coimbra and National Fire Brigade and Civil Protection Service (2005)

Figure 11.6 shows that fire spots were detected with a minimal value of relative humidity, but the number of detected fire spots was greater between the values of 50 per cent and 79 per cent of minimal value of relative humidity.

Figure 11.7 shows evidence that there is a tendency for a reduction in the number of fire spots when the values of minimal relative humidity are higher.

4. The urban–forest interface

There are different terms to name the fires that occurred near the urban areas: fires in the interface, in urban areas, in the urban–forest interface, or in the urban–wildland interface.

The urban–forest interface is a very vulnerable area in Portugal. In this area, 116 houses were burned in 2005, consisting of 59 primary residences and 57 secondary residences. There was also damage to 221 houses (181 primary and 30 secondary residences) and serious erosion effects after the fires in this area.

Butler (1974) uses and defines the concept of urban–forest as any spot where the fire fuel is artificial or made by humans (houses, annexes, etc.) and not natural (trees, shrubs or grass). When a fire occurs very near a housing area it may cause the house to burn due to the fire's proximity or because it emits burning particles.

Living in the wildland–urban interface means living with fire. Often there are fires in wildland–urban interfaces, and damage to property can be much greater than in forest areas. It is very important to educate and inform people living in forest or grass-land areas about precautions they can take to keep forest fires away from houses.

During the twentieth century, despite the fact that the number of people living in rural areas decreased, there were increasing numbers of people wishing to spend vacations in tourist areas located in the wildland-urban interface, to live in contact with nature, and in peaceful surroundings (Catarino 2003). In these areas, many people have built houses close to the forest without implementing measures to prevent fires.

The migration of the population from rural to urban areas has meant the abandonment of houses in rural areas, the growth of cities near forest areas, the abandonment of traditional agriculture (mainly in the last few decades), an increase in the number of fields that are not cultivated, and an increase in the quantity of wood in the forest that is not considered as the main fuel used for heating and cooking. The quantity of biomass and of fuel (any dead or living material that will burn) that remains in the forest therefore increases and becomes a source of ignition to spread fires in forest areas.

In Portugal, as noted above, the forest area has increased (Rego 2003) in the last few decades because of the reforestation of agricultural land mostly by species of eucalypt. There are now many houses and new urban developments located in forest areas. The result is a landscape where houses and trees are in close proximity.

The urban–forest interface is a pleasant place to live, but it is extremely dangerous for the habitat and for those who live in it (Carrega 1992).

In this urban–forest interface there are many secondary homes or houses belonging to emigrants where the yards and the gardens, as well as some agricultural land, are slowly being invaded by natural regeneration, gradually transforming grass and shrubs to mature trees, increasing the fuel mass near the houses and consequently the danger of fires.

Catarino (2003) considers two types of interface between forest areas and housing areas:

- compact residential areas that directly meet the forest, where many houses can be affected by a forest fire. These are usually small villages located inside forest areas
- residential areas or dispersed housing mixed with the forest vegetation, where there are isolated houses very vulnerable to fire. There are different types of buildings in these areas including tourist developments, and primary and secondary homes.

There are three main factors that cause the ignition of the houses:

1   The **kind of fuel** that is present in the area near the house. This includes the dominant species, the age of the vegetation, the fuel mass and its moistness. When a fire occurs and the dominant fuels are shrubs, there is a smaller possibility of the emission of particles and, consequently, the burning of houses is caused by radiation or direct contact with the flames. When the fire's available fuel is branches from trees, such as eucalypts, there is a greater probability of flammable particle emission that will accumulate in the roof of a house.
2   The **wind and the relief slope**. Wind influences the height and the length of the flames and the probability of the emission of burning particles. If the wind velocity is low the ignition of houses caused by particle projection is small. If the land is steep, the height and length of the flames is increased and consequently there is a high possibility of house ignition.
3   The **type and age of the buildings**. The type of building material is a very important aspect, because the ignition of a house is easier if the material is wood rather than, say, concrete. Older houses were usually built with more inflammable materials.

In the urban–forest interface, the following factors should be taken into account in evaluating the occurrence of a fire (Pita et al. 2005).

- **Road characteristics**: Density, width, good pavement conditions, and good visibility reduce the fire danger.
- **Vegetation**: A large quantity of thin fuel or dead material is more inflammable and ignitable.

- **Land slope and the location of houses.** The slope and the orientation of the hillsides influence fire behaviour (Catarino 2003). Usually when fires follow very steep slopes, the fire spread is quicker, and the secure area around houses must therefore be larger.
- **Building material of houses.** Some materials are more inflammable than others (see above).
- **The proximity of fire brigade and of water reservoirs.** This influences the time needed to combat the fires.
- **Urban infrastructures or equipment.** These can influence fire ignition.

5. Measures to tackle forest fires at the urban–forest interface

- **Legislation.** In Portugal there is legislation to prevent fires spreading. For example, the law prescribes that in rural areas administrative authorities are obliged to clear vegetation around houses and other buildings in a circle zone with the minimum distance of 50 metres. This increases to 100 metres in the case of villages and industrial areas adjacent to a forest. The protection zone is very important as the intensity and the speed of spread of a forest fire can be reduced.
- **Fuel management.** There are no species resistant to fires, but oaks, willows and weeping willows are more resistant than other species because they have high moisture and consequently they reduce the speed and spread of the forest fire. The large accumulation of dead fuel, dead trees, dry branches, deadfall or species with low moisture in summer (and which are therefore especially inflammable) or with oil or resin should be removed from residential areas.
- **Planning.** This might include implementing restrictions or guidelines to prevent the spread of forest fires, especially in the zones around houses, reducing the quantity of fuel and removing branches and deadfall that can fuel a fire. Cutting trees to increase the distance between them, especially on steep hillsides, prevents the fire spreading from tree crown to tree crown. Also, cleaning up the vegetation around houses (by either manual or mechanical means) and controlled fires (burn-off) has positive effects.
- **Development.** This requires the setting of guidelines for builders and developers to make developments less vulnerable to fire.
- **Training.** This includes the cross-training of firefighters who tackle both forest fires and house fires.
- **Interagency cooperation.** Municipal volunteer fire departments should work together with forest firefighting resources.
- **Education.** People living in forest or grassland areas should be educated about the precautions they can take to avert fires.

6. The freguesia of Ceira (Concelho de Coimbra) and the forest fires of 2005

6.1 Background to the fire

Ceira is a freguesia located in the centre of Portugal, in the southern suburbs of Coimbra, and is one of the most important Portuguese cities. Ceira occupies an area of 12.77 square kilometres, has a population of 4199 inhabitants (2001 census), and a density of 328.8 inhabitants per square kilometre. The freguesia is located on Massif Hespérico and the largest part of its territory is formed by rocks of the Complex xisto-grauváquivo (schist and grauvaques). The terrain is irregular with steep slopes in some places. The flat areas are located on the margins of the rivers Ceira and Dueça, which are respectively tributaries and subtributaries of the river Mondego, that cross the freguesia de Ceira. Ceira has an important historical heritage and unique geological features. People in the area work mainly in commerce, service industries, industry and subsistence agriculture. Around the Ceira and Dueça rivers there is a market agriculture based on nursery plants. Many people work in Coimbra, due to easy access to transport to the city centre. In the freguesia of Ceira there are several settlements, including Ceira, Vendas, Sobral, Boiça, Cabouco, São Frutuoso, Lagoas, Tapada and Carvalho.

The fire that began on 19 August 2005 affected a large part of the county of Coimbra (Concelho de Coimbra), including the freguesia of Ceira, and reached the school where this project was developed. It started in the neighbouring county of Vila Nova de Poiares. According to the General Department of Forest Resources (Direcção-Geral dos Recursos Florestais) the fire was caused by machinery and equipment. The fire started mid-afternoon and quickly spread from the county of Vila Nova de Poiares to the county of Coimbra and during the night the fire reached the freguesia of Ceira. The fire then crossed the river Mondego near the place where the school is located and reached the city of Coimbra. In the freguesia of Ceira the fire burned the majority of *Pinus pinaster* and eucalypt plantations. In the school, it burned the grassland and the trees of the school grounds. The total burnt area was 12 147 hectares and it was the biggest fire in the Coimbra district in 2005. In the same year, the district of Coimbra registered the largest burnt area in Portugal.

On the day the fire started—19 August 2005—the highest maximal temperature registered by the Geophysical Institute of the University of Coimbra was 23°C and the minimum relative humidity was 64 per cent. The ignition took place at the time of the highest temperature in the day (though it was not a day with a very high maximal temperature) and it was only during the night that the fire reached the freguesia of Ceira.

The fire showed evidence that urban–forest interface planning did not exist. Several houses were partially damaged and others slightly damaged. The circle zone, with a minimal distance of 50 metres of cleared vegetation around the houses, did not exist in the majority of cases. People did not choose the appropriate species of trees to grow near their houses, either because they were unaware of the differences between species in relation to the possibilities of ignition, or because they did not pay sufficient attention to fire prevention. A few decades ago traditional agricultural methods were used, which included cutting the undergrowth in order to make garden mulch (the undergrowth was mixed with cow excrement to make the mulch) so consequently the forest was periodically cleaned and trees branches were cut to use as firewood for heating and cooking. This way of living disappeared and the forest stopped being cleared. The population did not think about the possibility of a major fire in Ceira and was not prepared to cope with forest fires. Nevertheless, in Sobral, the fire brigade did not attend because there were many simultaneous fire spots. The local population used hoses to take water from the wells to extinguish the fire.

6.2 Some post-fire erosion effects in the freguesia of Ceira

The role of vegetation is important in many areas (but not always all) to prevent mass movement of slopes. Some months after the fire there was evidence of mass movement due to the fact that the vegetation was almost entirely burnt (especially the undergrowth and the shrubs). Vegetation usually reduces superficial erosion, because deep roots bind inferior layers and support the soil cover, infiltration is smaller, water evaporation is bigger and the vegetation (branches and leaves) reduces the superficial flow. Soil is not protected by vegetation after a fire and consequently there is serious soil erosion (see Figure 11.8, colour section, opp. p. 224).

On slopes, rocks of a few decimetres are dragged by water. Before the fire, slopes were covered by vegetation that blocked mass movement of soil and rock. In steeper areas there are examples of large blocks being dragged by water during the last winter season of 2005–2006.

The fire also had an effect on agricultural activity. In some areas there are terraces, covered with fine material with plenty of organic matter, where the infiltration is fairly large. Figures 11.9 and 11.10 (colour section, opp. p. 224) illustrate a terrace that was damaged during the 2005–2006 winter. This problem had several causes, including the fire that devastated the vegetation on the terrace, the abandonment of traditional agricultural methods which had previously preserved terraces, and the conduct of superficial water to canals through this terrace which increased the quantity of running water and contributed to its destruction.

For many years agriculture was carried out in certain sectors of the slopes, and the soil was consequently altered, with little debris cover. The soil was thin and not compacted, leading to conditions susceptible to erosion.

Some years ago the slopes were covered by *Pinus pinaster* and there was dense undergrowth with heath, strawberry trees (*Arbutus unedo*), butcher's-broom and other plants. Now the areas are covered with eucalypt species, and the plants formerly found there are no longer present. Consequently, when it rains heavily the run-off changes quickly from diffuse to laminar, and then to a concentration of water run-off. The water flows so strongly in streams that there is an erosive power able to carve deep ruts in the uncovered land (see Fig 11.11, col. sect., opp. p. 224).

All the aspects discussed above show the need to promote sustainable development in areas where the economy is changing quickly.

In Portugal, traditional agriculture (where the adaptation of human activities to the environment was reached after hundreds of years of development) is disappearing, or has already disappeared. The option of developing the forest areas in Portugal—taking into account the climatic conditions of the territory—without careful use of the forest raises huge problems because of forest fires in the dry season. The landscape is now hugely disrupted by human activities, which means that a sustainable development system has not yet been accomplished.

Forest fires have serious consequences, which include:

- environmental consequences, through land erosion, loss of plants and animals, degradation of the landscape
- economic consequences, as wood is an important source of income for the population and it disappears with fires, and erosion reduces the quality of agricultural land. There is often a loss of, or damage to, residences, equipment and even of industrial and commercial areas and tourist buildings
- social consequences, as people abandon rural areas, and contribute to the desertification of these areas.

Evidence has already been shown that there is a need for adequate legislation, planning, fuel management, firefighter training and education of people living in forest or grassland areas about the precautions they can take in dealing with forests and fire. A project on the forest fires issue has been developed in a school and the authors of this chapter hope to disseminate it to other schools throughout Portugal.

7. The Ceira School project

7.1 Background to the Ceira School

The school where the forest fires project took place is the Ceira School (Escola Básica 2,3 de Ceira). It is a middle school located in the central area of Portugal in the freguesia of Ceira, in the District and County of Coimbra (Distrito e Concelho de Coimbra). The school is located on a hill overlooking the confluence of the Mondego River and the Ceira River.

On the hill slope behind the school buildings there is a forest area that ends near the school grounds. On 19 August 2005 the forest fire reached the school grounds. Fortunately pupils were on vacation at the time and only the headmaster was in the school. He called the fire brigade and the school was saved.

Ceira School is a middle school (escola básica) with pupils from levels 5 to 9 (ages 10 to 15). It belongs to a group of schools that includes kindergartens and primary education schools (jardins infantis e escolas básicas do 1° ciclo). Ceira School had 224 pupils in the school year 2003–2004, 216 pupils in the school year 2004–2005 and 220 in the school year 2005–2006. Some children in the area prefer to study at other schools due to the difficulty of transport from some localities of the freguesia to the school, or because they prefer to study in the bigger city of Coimbra where their parents work.

Parents are mostly middle class, with the majority having academic qualifications from year 9, 4 or 6 of schooling. There is a much smaller number with higher or technical education qualifications. The majority work in commerce and service industries because Ceira is located in the immediate suburbs of Coimbra, one of the most important Portuguese cities. About 5 per cent of parents are unemployed and about 3.5 per cent are retired.

There are 58 teachers in the school, with 25 other people working in the school as psychologists, career guidance counsellors, administrative and auxiliary staff. The school is integrated in the Continuing Training Centre of Calhabé (Centro de Formação do Calhabé) which organises training for teachers and other staff in accordance with the needs of the school and its pupils.

The school building dates from 1996. It is an attractive building with outside grounds, good classrooms and a very good resource centre with library, computer and video facilities. However, the school does not have a physical education building.

In accordance with its Education Project (Projecto Educativo), the school gives special importance to:

- the socioeducational aspects of the curriculum, namely integration and participation in school activities by all members of the educational community
- the diversification and flexibility of solutions to educational problems, in accordance with school realities and the socioeconomic and cultural characteristics of pupils
- education based on equity for all
- quality schooling which prepares pupils to respect the future through human rights, their own initiative, the initiative of others, the natural and built heritage, the local and national identity, and the need for solidarity and cooperation.

This school was chosen to develop a project on forest fires for the following reasons:

- the location of the school in an area of rapid social, economic, environmental and cultural change
- the importance that the fires issue gained in the region in 2005
- the good organisation of the school
- the support of the school principal and the teachers involved in the project
- the collaboration of a colleague from the University of Coimbra, born and living in Ceira (Sobral).

7.2 Project rationale

In many countries, education for sustainable development (ESD) and citizenship education (CE) are considered to be priority areas, both at the political level and among those who work in the fields of education, social cohesion and sustainability. Many initiatives, such as the UN Decade of Education for Sustainable Development, various European conferences and curriculum legislation in many countries, reflect a sense of urgency in this area.

To meet the needs of ESD and CE, school curricula should include the acquisition of knowledge and understanding about environmental, cultural and development aspects. This knowledge should be specifically linked to the concepts of:

- biological and cultural diversity
- globalisation and interdependence
- sustainable development
- peace and conflict
- social justice and equity.

Understanding these concepts is fundamental when engaging pupils in sustainable development actions from local to global levels. There is also a need to organise activities that promote the development of skills such as:

- critical thinking
- arguing effectively
- challenging injustice and inequality
- respecting the environment and people's cultures
- cooperation
- problem-solving and decision-making.

Pupils should also develop values and attitudes, such as a sense of identity and self-esteem, empathy, commitment to social justice and equity, commitment to sustainable development and belief that people can make a difference (Ferreira 2002; Miranda, Alexandre & Ferreira 2004).

The above principles formed the basis of the Ceira School project's organisation.

Project aims

The project's overall aims were to:

- contribute to education for sustainability and active citizenship; namely, to the development of pupils' concepts of globalisation, population mobility and sustainable development, and to the development of pupils' commitment to an active participation in the resolution of sustainability issues
- raise awareness in the local population of the need to actively participate in the prevention of forest fires.

Project objectives

The project's specific objectives were to:

- help pupils acquire a deeper knowledge and understanding of the causes of forest fires (that is, linking fires with climate characteristics and human activities)
- help pupils to acquire a deeper knowledge and understanding of the environmental, social and economic consequences of forest fires
- develop pupils' sense of enquiry, and their group work and problem-solving competencies
- develop pupils' capacities for active participation in forest protection
- develop pupils' concept of sustainable development and attitudes in relation to active participation in sustainable development issues.

Expected outcomes

The expected outcomes were:

- development of pupils' knowledge about the concept of sustainable development and active citizenship
- development of pupils' active participation in the resolution of local issues specifically related to forest fires
- development of pupils' competencies in enquiry methods and problem-solving
- development of active participation by the local population in the resolution of local issues in collaboration with local authorities
- presentation of the results of the research, and public discussion at a local level.

Target group

The project was developed at school level with pupils from the seventh form (ages 10–12).

Key people

School teachers, school directors, the local population and local authorities were involved.

Local political context relevant to the project's success
Active collaboration of local educational and administrative authorities was
necessary for success.

Project activities
The tasks for the students were to:
- use data already collected by government and local authorities for mapping the
  fire areas
- analyse the relationship between meteorological conditions and the ignition of
  fires
- analyse the conditions at the local level of the urban–forest interface
- interview local people and local authorities to collect data about the causes and
  consequences of fires
- undertake fieldwork to collect data (including photographs) on the conditions
  of forest now (after the 19 August 2005 fire) and of erosion problems
- form groups to analyse data, and to discuss results and findings
- present the results of group work and do an exercise on problem-solving
- organise an exhibition of the results of the enquiry and invite the local population
  and local authorities to a joint meeting
- send the results to regional and central authorities and to the local newspapers.

Project evaluation
The project would be evaluated by:
- assessing group work, the quantity and quality of collected data, the treatment
  and results
- evaluating the quality of the enquiry results through a questionnaire administered
  to visitors to the exhibition prepared by pupils
- involving the local population and local authorities to the joint meeting.

Evaluation criteria
- How transferable is the project to other countries, and to regional and local
  contexts?
- How does the project contribute to a wider understanding of sustainability?
- Does the project allow for translation into multiple languages?
- Does the project contribute to building communities of practice?

Summary of developed activities
- The school year in Portugal starts in September.
- Pupils have already collected, treated and analysed data on burnt areas, and the
  causes and consequences of forest fires.
- The project was developed until the end of the school year in June 2006.

- In order to deepen pupils' knowledge of forest fires, one of the authors of this chapter (Jorge Duarte) will work with pupils on satellite images to assess the impact of fires in Portugal, meteorological conditions favourable to the ignition and spread of forest fires, fires in the urban–forest interface, and the effects of fire on erosion.

8. Conclusions

In Portugal, forest fires hinder sustainable development. According to UNESCO, education for sustainable development (ESD):

- *deals with the well being of all three realms of sustainability—environment, society and economy; …*
- *is locally relevant and culturally appropriate;*
- *is based on local needs, perceptions and conditions, but acknowledges that fulfilling local needs often has international effects and consequences;*
- *engages formal, non-formal and informal education; …*
- *addresses content, taking into account context, global issues and local priorities;*
- *builds civil capacity for community-based decision-making, social tolerance, environmental stewardship, adaptable workforce and quality of life;*
- *is interdisciplinary. No one discipline can claim ESD for its own, but all disciplines can contribute to ESD, and*
- *uses a variety of pedagogical techniques that promote participatory learning and higher-order thinking skills. (UNESCO 2005)*

In this chapter, an important issue was illustrated at national, regional and local levels. The issue of forest fires involves the three realms of sustainability quoted above (environment, society and economy), as well as cultural aspects. The project at Ceira School is based on local needs and is being developed within the framework of formal education, but it also aims to reach the entire local population to enhance community-based decision-making and environmental stewardship. The project is interdisciplinary, although geography has a central role in it. A variety of pedagogical techniques have been used and an expected outcome is the development of pupils' thinking skills, as well as the development of respect for the environment, taking into account the rights of the present and future generations.

Bibliography

Butler, CP 1974, 'The urban/wildland fire interface', *Proceedings of western states section/ Combustion Institute papers*, Washington State University, Pullman, WA, vol. 74, no. 15, pp. 1–17.

Carrega, P 1992, 'Risque de feu de forêt et habitat dispersé dans le Sud de la France', *Finisterra*, vol. XXVII, nos. 53–54, pp. 154–78.

Catarino, V 2003, 'Floresta e incêndios', *Revista técnica e formativa da Escola Nacional de Bombeiros*, 26, vol. 7, pp. 7–16.

Direcção-Geral das Florestas 2003, *Incêndios florestais 2003*, DGRF, Lisboa.

Direcção-Geral dos Recursos Florestais, DSDF–Divisão de Defesa da Floresta contra Incêndios 2005 *Incêndios florestais 2005—Relatório Provisório 801 Janeiro a 09 Outubro*, DGRF, Lisboa.

Ferreira, M 2002, 'Environment and citizenship: From the local to the global', *Geography, culture and education*, Kluwer Academic Publishers, Dordrecht, Netherlands.

Lourenço, L 1991, 'Aspectos sócio-económicos dos incêndios florestais em Portugal', *Biblos* vol. 67, pp. 373–385.

Miranda, B, Alexandre, F & Ferreira, M (eds) 2004, *Sustainable development and intercultural sensitivity: New approaches for a better world*, Universidade Aberta, Lisboa.

National Fire Brigade and Civil Protection Service 2005, *Autoridade Nacional de Protecção Civil*, ANPC, Portugal, viewed 30 July 2007, (http://www.proteccaocivil.pt/Pages/default.aspx).

Pita, L, Cruz, M, Ribeiro, L, Palheiro, P & Viegas, D 2005, *Curso sobre comportamento do fogo florestal e segurança das populações*, Associação para o Desenvolvimento da Aerodinâmica Industrial Coimbra.

Rebelo, F 2003, *Riscos naturais e acção antrópica—estudos e reflexões*, Imprensa da Universidade de Coimbra (2ª edição revista e aumentada), Coimbra.

Rego, C 2003, 'As florestas portuguesas', *Ambiente 21—sociedade e desenvolvimento*, 11, ano II, pp. 12–28.

UNESCO 2005, *Draft International Implementation Scheme (IIS) for the United Nations Decade of Education for Sustainable Development (2005–2014)*, UNESCO, Geneva, viewed 5 June 2007, (http://unesdoc.unesco.org/images/0014/001403/140372e.pdf).

Notes

1   Decreto-regulamentar N.º 5/2004, DR n.º 94, I-B Série, de 2004.04.21, Ministério da Agricultura, Desenvolvimento Rural e Pescas.
2   Continental Portugal is divided in 18 Districts (*Distritos*). The District of Coimbra is divided into 17 Counties (*Concelhos*). One of these counties is the County (*Concelho*) of Coimbra (in which is situated the city of Coimbra). This is further divided into 31 *freguesias*. Ceira is one of the freguesias of the Concelho de Coimbra.

12 | Geographical perspective on training of students in sustainable development in Georgia

Niko Beruchashvili

1. Georgia

Georgia is a small country of about 70 000 square kilometres, located on the border of Europe and Asia, though Georgians consider themselves to be European and wish to be integrated into Europe. Georgia is located on the Caucasus, the highlands between the Black Sea and the Caspian Sea. In the north, Georgia shares its borders with the Russian Federation, in the south with Turkey and Armenia, and in the east with Azerbaijan. West Georgia is washed by the Black Sea.

The population of Georgia in 1989 was 5.6 million but in 2006 was only 4.4 million. This sharp population reduction is the result of many inhabitants migrating to other countries in search of work, and also a low natural population increase. The death rate currently exceeds the birth rate. The capital of Georgia is Tbilisi, with a population of a little over one million.

Georgia possesses an extraordinarily varied set of environments, from lowlands to almost impassable mountains. The highest peak, Shkara, is 5200 metres. The climate is also varied, with damp subtropical conditions in the west (up to 4000 millimetres of precipitation per year) and a dry climate in the east (from 200–500 millimetres). There are even fragments of deserts, and high mountain conditions in the Caucasus.

The flora and vegetation of Georgia is also extremely varied. In the west there are relics of Colchic forests which are similar to tropical forests in character. In mountain areas there are many oak, beech and fir forests. Forests cover up to 40 per cent of the country. In the east, steppe conditions are found, and in high mountain areas magnificent alpine and sub-alpine meadows are widespread. Above 3000–4000 metres, mountain glaciation occurs in significant areas of the Caucasus.

Georgia's economy is mainly agricultural. Georgian wines are known all over the world, and in the past Georgia supplied the Soviet Union with citrus fruits (tangerines, oranges and lemons) and tea.

During the years of Soviet Union administration, Georgia had well-developed industries, but now practically none remain. In the past, one important income source was tourism. Georgia was visited by up to five million tourists from different areas of the Soviet Union, but now the greatly reduced tourist stream consists only of locals or tourists from Armenia.

Georgia completely depends on external power resources and although nowadays the power problem is practically solved, any failure of gas or electricity supply can lead to an economic and social collapse. A vivid example of this occurred in late January 2006 when gas from Russia was disconnected, causing a shutdown of electricity to the country and leaving parts of Georgia without heat or power for two weeks.

2. Georgia and the geographical perspective of sustainable development

In Georgia, many geographical problems are connected with sustainable development. Three particular problems (or themes) were chosen for training students in universities and high schools. All three problems were examined during a meeting in Rome, 2005.

The first theme to be studied was the problem of restoring Tbilisi National Park. Fifty-five students of geography at the Tbilisi State University were organised to participate in a study of sustainable development. The main task of the students was to understand that there are alternative ways to achieve sustainable development and these have been solved by concrete research in the field. Researchers spent two weeks at the geographical station of the Tbilisi State University in Martkopi. The practice was divided into three stages. First, students worked together with teachers and were trained in the methodology of measurement-taking necessary for understanding the problems of sustainable development. At the second stage students did independent research. At the third stage, the received results were processed. The study ended with a round-table debate. During discussions, all students were divided into three groups. These groups represented the interests of foresters, ecologists and the local population. The results of the study are discussed below.

The second theme concerned virgin landscapes in Georgia, and special lectures for students of the Tbilisi State University were devoted to it.

The third theme was connected with the problem of estimating critical territories in the landscape. Three lectures for undergraduates of the university were devoted to this theme.

From the three themes, only the work of students connected with the first theme was fully developed and investigated, as shown below.

3. The Tbilisi National Park

3.1 Background

In January 1973, in response to the Stockholm Conference of the United Nations, the USSR published the paper *About improvement of wildlife management*. In March 1973, based on that publication, Georgia accepted the decisions and published *About expansion of a green zone, improvement of an environment and improvement of conditions of mass rest of workers in the cities of Tbilisi and Rustavi*. Forest areas in Gldani, Martkopi, Gulelebi and Tskvarichamia were defined as the Tbilisi National Park and the borders were confirmed on 1 January 1974, with the purpose of:

- protecting the important objects and areas within the park, its rare plants and animals, and historical and cultural monuments having architectural and ethnographic value
- creating the base necessary for scientific research and attracting cultural–scientific institutes
- popularising and acquainting visitors with the park.

A wildlife reserve in the park was created, centred on wildlife management, tourism and research work.

The creation of the park demanded the establishment of a network of roads, a system for wildlife management and the allocation of recreational areas. The problem was to define which parts of the park's territory should be allocated for recreation. It was accepted that the area should be 10 per cent of the total park, should be away from the most valuable and sensitive parts of the forest, and should be close to roads.

Tbilisi National park is located north-east of Tbilisi, and it includes the forest parks of Gldani, Martkopi, Tskvarichamia, Gulelebi, and Mamkoda Memorial Park. The territory of the Tbilisi National Park is located between 41°30' and 42°30' north longitude, and 44°40' and 45°07' east longitude. The total area is 1913 hectares, with the nearest part 10 kilometres from Tbilisi, and the greatest 41 kilometres.

The Tbilisi National Park is located in three administrative districts—Tianeti, Mtskheta and Gardabani—and includes seven timber enterprises. The territory of the Tbilisi National Park is in the area of Gldani, only 10 kilometres from the administrative border of Tbilisi. The most remote part is in the area near the village of Didi Lilo. In the north-west Tbilisi National Park shares borders with the wildlife reserve at Saguramo. In this reserve there is a strict code of wildlife management and only visitors with special scientific passes are allowed. The significant part of the territory of the Tbilisi National Park is within the vicinity of the Martkopi geographical station. This area is very well studied both in geographic and land-scape terms.

The park is 19 134 hectares in size, comprising forest (18 374 hectares) and non-forest areas (760 hectares).

Within the park there are about 50 kinds of trees and bushes, including 18 kinds of Colchic plants, which make up 36 per cent of all dendro-flora in the national park. In the areas covered by forest, the oak, beech and hornbeam prevail, and they occupy 16 122 hectares (92.8 per cent) of the area. The relative density of coniferous (pine) forest stands is low at just 1.5 per cent, with middle forest stands prevailing at 57.4 per cent and old forest at 16.8 per cent.

Attention was given to the functional zoning of the territory of the national park, and four basic zones were determined, comprising wildlife reserve, productive leisure, 'silent' areas, and exploitation zones.

- The **wildlife reserve** was allocated only 4574 hectares (24 per cent) of the total park area, with vertical ash being well represented on Ialno Ridge. Economic activities, except for necessary scientific works, have been forbidden, but entry to the wildlife reserve is permissible by groups on excursions. Unfortunately, the plans have not been realised and the establishment of the wildlife reserve in the Tbilisi National Park has not taken place.
- The **zone of productive leisure** comprises 4420 hectares (24 per cent) of the total park area. The area requirements include sanitary and hygienic conditions, aesthetic sites, forest landscapes, places rich with architectural monuments, and places with flower and fruit crops. In this zone the construction of campsites, vacation spots, picnic areas with tables and cafés, etc., was proposed, the construction of which should be completed in an environmentally sensitive way. This territory of the park could accept about 19 400 leisure-seekers.
- The **zone of silent rest** comprises 8025 hectares (43 per cent) of the total territory of the park. The area requirements include places for silence and rest, and for aesthetic education.
- The **forest exploitation zone**, of 2125 hectares (9 per cent) is not intended for rest and recreation but for forest exploitation, with the purpose of improving the condition of the forests and their restoration. These activities will depend on the current condition of the forest.

3.2 Closing the Tbilisi National Park

In 1991 the Soviet Union as a single entity was broken up. Abkhazia and Ossetia, former independent units of Georgia, did not want to remain within the structure of the new Republic of Georgia. War began in Ossetia and in Abkhazia, and then civil war in Georgia. The economy of Georgia declined, and the country lost a source of raw materials (especially oil and gas) from the former USSR. For 80 years forests in Georgia had not been cut for industrial purposes or for household use, but the delivery of timber from Russia stopped and this has led to illegal cutting in forest areas.

Though Georgia is located on the border of the subtropical and moderate zone, temperatures often fall in winter months to below freezing. The population of Georgia, especially in winter, endures severe conditions, often without electricity or gas for heating or cooking. In Tbilisi the population cut down trees in streets, cemeteries and parks to use as a source of fuel. In the Tbilisi National Park, with its good forest resources, illegal cutting of the forest began. The management of the park was unable to protect its borders, and nothing remains of wildlife management. Therefore in 1995 the Tbilisi National Park was closed after being in operation for 21 years.

3.3 The basic stages in the development of the park

- **1973–84:** The Tbilisi National Park was created in 1973 and infrastructure, including roads, began to be developed. Unfortunately, development was conducted in an uncontrolled manner, and along the new roads areas of forest 50 metres wide were cut. Access to sources of potable water was also investigated at this time, as these sources were necessary to attract tourists. During this period the network of roads became gradually degraded because they were not being used. The villagers of Martkopi and Norio started to use kerosene furnaces (kerosene during this period was very cheap) so the demand for wood fell. It would normally have been taken out on bullocks and horses along a densely branched network of forest roads but these fell into disuse and a forest renewal period began.

- **1985–91:** This was the period of active construction of summer residences in the vicinity of Tbilisi. Martkopi became a housing estate, and was supplied with gas for fuel, so a decreased demand for forest products occurred in this period as well.

- **1992–95:** This was the period of deepest economic crisis. To distract peasants from active political agitation, the Government of Georgia declared that land would be privatised and distributed. The creation of small farms, gardens and kitchen gardens began. Many townspeople, because of economic ruin and conditions in the city, came back to the villages and started to engage in agricultural activity. At the beginning of this period the demand for forest products was still small. However, in 1995 the government made the decision to close Tbilisi National Park so it became unsupervised and open to active penetration by poachers.

- **1996–2003:** This period was one of stagnation. Though the economy was stabilised, the energy crisis continued. Supply of gas to the villages stopped, so peasants cut firewood from the forest not only for local consumption, but also for use in city furnaces. The sale of firewood became a lucrative business, with a cubic metre of wood selling for the equivalent of about $20. Foresters autocratically changed the park's status. They put numerous cordons at the entrance to forests and supervised peasants who cut wood. Peasants bought special permits for cutting set quantities of wood, or they bribed foresters, who subsequently grew rich. For many peasants the sale of wood became a good source of income. The forest saved a significant number of peasants from ruin, but rich native forest areas were reduced as a result of illegal cutting. Farming and kitchen gardens were also abandoned as the peasants did not have the machinery to harvest.

- 2003–05: The energy crisis has been partially overcome as gas supplies to the villages of Martkopi and Norio have been restored. The need for forest products decreased again.

3.4 Restoring the park

At the time of writing, the economic situation in Georgia has improved. The energy crisis is mostly resolved, and gas flows from Russia, Azerbaijan and Iran. The restoration of the Tbilisi National Park is again an issue. Its problems are the same as those when the park opened in 1973. However it is also necessary to consider the negative experiences in the park's creation. First, precise landscape planning on a landscape-ecological basis should be done, similar to the numerous research studies made at the Martkopi geographical station. Second, it is necessary to think how zones of recreation in the national park will be organised, and to consider the interests of the users. For example, should there be some areas for amateur bushwalkers, and different ones for, say, excursion tourists or researchers? Third, the interests of the local population should be considered by leaving some forest territory for pasture and haymaking. Finally, the organisation of the park should be considered, especially within those territories which have recreational or scientific value but are not places of intensive economic activity.

It is necessary to restore the Tbilisi National Park, but this restoration should occur on the basis of precise scientific planning, taking into account those mistakes made from 1973–1995.

4. The Martkopi geographical station of Tbilisi State University

The Martkopi geographical station of Tbilisi State University was founded in 1965. The basic purpose of the station is to carry out fieldwork and long-term research into the conditions of the area.

The station operates all-year round, and takes about 5000 to 6000 data readouts per day. More than 100 parameters describe daily conditions of the structure and function of the complex natural territory. There have been 150 articles and 10 monographs published based on this data. Three employees of the station have completed doctoral degrees and 10 have Masters qualifications.

One major function of the Martkopi station is the carrying out of field practice for geography students. Over 3000 Georgian students and 450 students from other countries have studied at the station. The Martkopi station is located 30 kilometres north-east of Tbilisi at a height of 910 metres above sea level, in foothills with steppe vegetation. Nearby is the 1800-metre Ialno Ridge, which is very convenient as students can observe the entire landscape characteristic for East Georgia (except for glaciers) during a one-day walking tour. In the foothills there is steppe vegetation and at a height of 1000 metres there are low mountains with oak forests. At 1400 metres this gives way to beech forests, and at 1700 metres there are magnificent sub-alpine meadows.

5. The project: Training students in sustainable development

Training students of geography about the practice of sustainable development took place in the summer of 2005, for a period of two weeks. Fifty-five students took part in two curriculum areas of the university. All student geographers worked with geography staff who have specialist skills in cartography, geoinformatics, physical geography, soil science, economic geography, hydrology and meteorology. The students ranged in age from 19 to 22, with 30 female and 25 male. Their educational levels (based on the results of the previous two years) ranged from average (60 per cent), to above average (33 per cent) and honours (7 per cent). The selected group is a representative sample of all Tbilisi students. The 55 students lived in two houses at the Martkopi station with five or six students per room. Professor Niko Beruchashvili led the project, assisted by senior lecturers Tenguiz Gordeziani, Robert Maglakelize, Temur Zirakishvili, Nili Jamaspashvili and Tenguize Dekanoidzw, and teacher Roman Maisuradze. Staff lived in a separate teaching billet. Four areas in the living space were assigned for laboratory measurements and processing of the received data, and also for joint lectures and conferences.

6. Strategic problems and approaches in the project

It was decided to break the practical parts of this project into three parts. During the first stage, students worked together with teachers. The second stage involved independent student work, while the third involved the processing of data material. The study ended with a round-table debate during which students discussed different ways of maintaining sustainable development of the territory.

6.1 The first phase

During the first part of the study there were general lectures in which the environment around Martkopi station and the general territory of the Tbilisi National Park were described. The main part was fieldwork with teachers, which took place in key parts of the park. During this time students became acquainted not only with the nature of the territory but also, under the direction of the teachers, mastered the elementary basics of geographic measurement.

Attention was especially given to the gathering of a herbarium, and storing many kinds of plants. At the end of this period a number of intermediate 'tests' were given. If a student could define 150–200 plants, maximum marks were given; if 100–150 the grade was 'good'; 50–100 was 'satisfactory'; less than 50 was 'unsatisfactory'.

Each student compiled a field diary to note observations made on the route, and record key points made by teachers in the field. Teachers assessed the detail, quality and cleanliness of the field diaries daily, and at the end of the study. Special attention was given to originality of thinking, self-observation rather than directed study, and analysis of the data obtained.

The results of the analysis of the field diaries corresponded to students' educational levels. Many students wrote poorly and could not clearly state what they had seen. Eighteen students did not complete a diary and used 'playing football' or 'sleeping' as an excuse. Only 16 students presented 'normal' diaries and from these only eight (six female and two male) were deemed 'good'.

One project involved drawing a point of the site based on coordinates received by GPS technology onto a topographical map. As it was concrete work, this created great interest in the 45 who participated.

A lot of attention was given to training students in the skill of decoding aerial photographs. Students at second-year level had already passed a theoretical and a practical test at university on work with aerial photographs, but experience showed that many students had forgotten their university training and found the exercise difficult when in field conditions. A lot of time was therefore given to the analysis of aerial photographs and comparisons with the actual situation of an area. During their first year at university, student geographers study topography and also master methods such as estimating areas by eye, but by the end of the second year many students have forgotten these skills. Therefore during the first phase of the study teachers revised previously learned skills.

Students filled in forms, describing an area where detailed research was carried out, under the supervision of teachers. They were shown how to answer the questions contained in the forms, and how to fill columns with the measurements recorded.

6.2 The second phase

The second part of the study was deemed the most important. Students were broken into four 'Brigades'. In each Brigade the student with good scientific knowledge and good organising abilities was nominated as team leader. These students were chosen from the eight people who had the best field diaries and had the best data on all parameters which were set at the first phase, including skills in plant knowledge, placing findings at the correct point on a topographical map, analysing aerial photographs, descriptive diary entries, filling in forms, interpreting maps and so on. The principal staff member, in coordination with other staff, distributed the remaining students into Brigades with approximately equal skill levels. Some students did not like this distribution, as they wanted to choose the groups for themselves, but it was pointed out that the allocations given were important for efficient study purposes.

Each Brigade was set the following task: find out which changes have occurred in the environment since the national park closed. Sets of topographical maps and aerial photographs from two different dates (1990 and 2002) were used.

This was not a simple task. Finding the precise locations of the old (1990) descriptions of areas proved difficult. Unfortunately, not all groups managed to find these places, and they often resorted to help from teachers. Once found, it was necessary to take measurements of all remaining trees, including the species, diameter, height and form. Special attention was given to measuring the diameter of the trees which had been cut down, and determining when they had been cut (see Figure 12.1, colour section, opp. p. 224). Over the past 10 years many stumps had been removed, so students used indirect attributes such as intensive grassy vegetation and the presence of moss spots to estimate how many trees had been cut. Unfortunately, this was too subjective and there were major variations from site to site and from Brigade to Brigade, so the data could not be used.

In total, each Brigade made 10 points of detailed field descriptions and a special form was designed for this. On this form the following data were entered.

- **The 'General' block:** This included a descriptive name of a site, using GPS coordinates. It was also important to find, using an old map of the Tbilisi National Park, the number of a quarter and a forest site. The number of a point of the described site was appropriated as follows: the first figure corresponded to the number of a Brigade, and the subsequent figures to a number of a point of the description in a Brigade. For example, '305' refers to '5 points by the third Brigade', and '210' to '10 points by the second Brigade'. The establishment of this was important in determining the described area and in compiling an index of a natural–territorial complex (defined after filling in the form, on the basis of the analysis of all components found).

- **The 'Natural' block:** This included the description of a relief and a geological structure—that is, the physical geographical description of a point. For this purpose, a list of the kinds of plants in the area was made, and a description of the area of vegetation, including its height and abundance. If students could not determine the name of a plant they took this plant to a herbarium and indexed it in an appropriate way, for example as '405–5', where '405' designated the number of a site, and '5' the number of uncertain kinds of plants. A teacher later identified the sample and recorded the details.

- **Research of an underground part of a landscape:** For researching underground parts of the landscape, such as soil or rock, a one-metre hole was dug and details of soil, etc., were described relative to ground level. Comparison with old descriptions (from 1990–95) have shown that some changes have occurred only where all trees have been cut down, and these changes have occurred only in the uppermost part of the ground, as 10–15 years is too short a time for ground change. In those places where the forest was partially kept, changes have not occurred or were within narrow limits of research error.

- **Definition of the quantity of a forest:** This was a very important stage in the research. The diameter of trees was defined by means of a forest plug, which is widely used in forestry. The height of trees and their defined form were estimated by eye, for which special training under the direction of teachers had been given to second-year students. Knowing these three parameters (diameter, height and form) it was possible to use special forest tables (Nikolsky's table) to determine the extent of forest at a site, and to transfer this into hectares.

- **Mapping natural territorial complexes (micro landscapes):** This was perhaps the most difficult area of study, especially mapping those changes to an environment which have occurred during the past 10 years. Not all Brigades coped with this analytical problem and consequently they often required the assistance of an experienced teacher.

6.3 The third phase

At the third stage, material was processed and the data analysed. This was the most interesting part for the students, especially seeing the analysis of their own measurements and comparing them with the results of the other Brigades. By this stage there was some stratification of Brigades into those students who actively worked on the project, and those who were more passive. Interestingly, this project helped students recognise whether their interests lay in geography or not. The male students were usually more active in the field, and practical activities were found to be more interesting than cramming subject matter before an examination. However, the two Brigades supervised by female students acquitted themselves very well.

Data were processed based on measurements made during fieldwork and the area of forests was calculated. Ground analysis was verified and the calculations were made using Excel spreadsheets. Each Brigade's data of the 10 points described were also transferred to Excel. Finally all described points were plotted. Two maps were produced. The first showed the quantity of stock in the forest, and the second showed changes to the environment since the park closed. Maps were made by using the computer program MapInfo.

The first and third Brigades gave fuller and more exact data, whereas the second and fourth Brigade's were less exact. The detailed map of changes to the environment of the territory of Tbilisi National Park was made with the help of the teachers.

At the final stage each Brigade completed a written report which came to about 30–40 pages, including schedules, maps, tables and photos.

7. The scientific results of the research

A special map was made showing the loss of forest areas due to cutting. Figure 12.2 shows the gorge area mapped by Brigade 1.

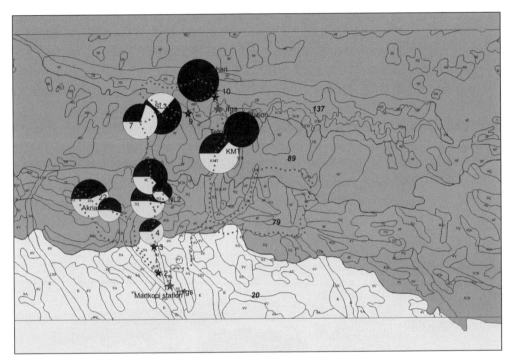

Figure 12.2   Cutting of forest—a map made by the first Brigade

The black part of the circle indicates forest stocks in 2005, and the lighter part the cut forest. The total circle area therefore corresponds to stocks of the forest at 1990–95 levels. For this period there was forest regrowth, and is especially visible at two sites in remote areas of the Ialno Ridge, on steep slopes. There are no roads here so the area was not reached by woodcutters. Calculations have shown that stocks of forest have increased about 7–10 per cent in the area, and are likely to be the same in similar areas. However these figures lay within the limits of measurement error and consequently may not be accurate.

Sites have lost from 20 to 70 per cent of their original forest area. Nevertheless, no sites were completely cut down, and this is connected to the psychology of the Georgian peasant. Centuries-old traditions have encouraged a careful relationship to forest areas. Usually only the most valuable species of trees were cut down, the most productive being beech. As a result beech forests have been replaced by hornbeam.

Most losses occurred along roads and in easily accessible sites. Tractors were used in some areas but even they could not reach the steeper slopes. Therefore there are some territories which have not been cut and have the character of virgin landscapes. Nakokhari, on the northern slope, the Ialno Ridge, Ialno, and Samebiskhevi, on a southern slope, are examples.

Figure 12.3 shows the degree of change of the landscape since the closing of the Tbilisi National Park. Five categories were defined:

1   **Big changes.** These are territories which, as a result of cutting, have lost more than 50 per cent of their forest stocks. Foothill forests located close to villages and roads and accessible by tractor were most affected.

2   **Moderate changes.** These are territories which have lost from 20 to 50 per cent of their forest stock. These occupy about 30 per cent of the area near Martkopi station and consist of low mountains with oak and beech forests.

3   **Weak changes.** These areas have usually less than 5 per cent decrease, and are found in remote mountains with beech forests. These are former collective farm forests, and fall into the same category as foothill zones which were not part of the Tbilisi National Park, and which during the Soviet period were intensively cut. They are also on the border of a dry climate ecosystem and it is known that

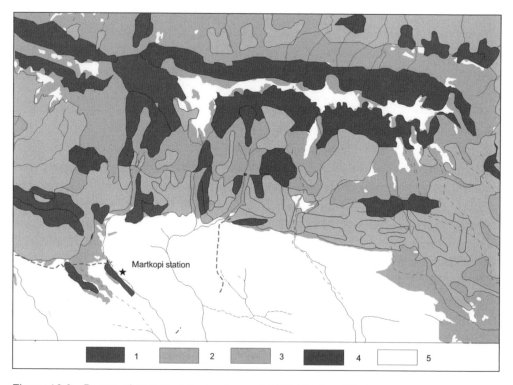

Figure 12.3   Degree of change of the landscape in Tbilisi National Park

400–500 millimetres of annual rainfall is required or the forest cannot grow. These figures are characteristic of east Georgia's hot climate.

4   **Practically untouched forests**. In these areas stocks of forest have increased. The area includes the steep slopes of the higher mountains and accounts for about 10 per cent of the total area.

5   **Territories without a forest**. These are foothills with agricultural land and villages.

8. Round-table discussion concerning sustainable development

On the final day of the study there was a round-table discussion concerning sustainable development. Students were required to address only one question: from the point of view of sustainable development, is restoration of the Tbilisi National Park necessary?

Students were broken into three groups: foresters, ecologists and local people. At the beginning, students protested about this division of groups. The majority of the 'active' students were already defined with a position and wished to remain in their specialist groups. Therefore, students were distributed in three unequal groups. The greatest number were in the ecologist group and included 18 students from the city of Tbilisi. The majority were female. The second group, the foresters, consisted of 15 students, mostly male. These students, generally of village origin, were in the practice groups rated 'average' or 'good'. There were no honours students in this group. Ten people formed the third group (local people). However, 12 students were not included in any group, as their practice experience was rated 'poor' or 'below average'. They agreed to contribute to the discussion only if asked for an opinion.

The ecologist group offered the following arguments in favour of restoring the Tbilisi National Park:

- a city of one million people requires its own national park
- the creation of a national park will lead to wildlife management in the vicinity of Tbilisi and to improvement of the ecology, both in the territory of the park, and in Tbilisi
- the creation of a national park will lead to a cessation of illegal cutting of the forest.

The foresters argued that:

- the cutting of the forest has rescued a significant part of the population from a full economic crisis
- the cutting of the forest remains a source of income for many peasants
- the cutting of the forest has not led to essential changes in the environment of the national park.

The ecologists countered with the argument that during the existence of the Tbilisi National Park there were 44 natural sources of water, and now there are only 24. The important argument is that the territory of the park is a complex ecosystem and essential for the supply of natural water, and should therefore not be tampered with. The foresters replied that although many water sources had disappeared it was not because of deforestation. The small quantity of water available always deterred fast development and the absence of water even during the Soviet period stopped many tourists from visiting.

Much discussion took place between the ecologists and the foresters while the group representing the local population behaved rather passively. However, they were asked: what will the creation of the Tbilisi National Park bring to the local population? They stated it was likely to increase the summer residence population like the boom seen from 1985–91, which might lead to the same infighting between businesses which had occurred earlier. The ecologists noted that the park would bring in tourists, benefiting the locals through the sale of food and souvenirs. The foresters objected, stating that the creation of more summer residences between villages and the forest will lead to further deforestation.

Discussion on the topic proceeded for three hours and was recorded. The following observations were made after the discussion:

- 'mediocre' students appeared to respond more actively during and after the study because of the concrete work involved
- female students strongly deferred to male students when understanding the purpose of the exercises, and in discussions
- the group representing the interests of the local population gave poor feedback
- discussion was mainly conducted between 'ecologists' and 'foresters'
- sympathies were on the side of ecologists
- students understood that there are variants of sustainable development
- foresters argued that essential changes have not occurred
- ecologists (sometimes speculatively) argued that there were irreversible changes, although they did not have access to factual data.

At the end of the discussion, the students came to the common opinion that restoration of Tbilisi National Park is necessary. However the country will need to develop higher economic standards; otherwise the Tbilisi National Park will be a park that only appears on paper, rather than in reality.

One of the major results of the practice was that students understood the concept of sustainable development. It became clear to them that:

- sustainable development is a concrete reality
- concrete field research is necessary for work on sustainable development
- results of field supervision and research should be processed and analysed to a high standard

- there can be different variants of sustainable development and that it is necessary to respect differing points of view.

Finally, students enjoyed their practice on sustainable development. Many students declared that while at school they will try to apply those methods in which they were trained. It is a very important result as more than 80 per cent of students of the Tbilisi University work in schools after completing their studies. Thus, questions of sustainable development will be included in the high school curriculum and will remain as a topic for serious discussion.

Bibliography

Beruchashvili, N 1989, *Landscape ethology*, Tbilisi University Press, Tbilisi.

—— 1990, *Landscape geophysics*, Vyschaia Skola, Moscow.

——, Chauke, M & Sánchez-Crispin, A 2004, *Geographical perspectives on sustainable development*, International Geographical Union, Rome/Moscow/Beijing.

—— & Zhuchkova, V 1997, *Methods for complex physical-geographical investigation*, Moscow University Press, Moscow.

Rougurie, G & Beroutchachvili, N 1991, *Geosystèmes et Paysages*, Armand Colin, Paris.

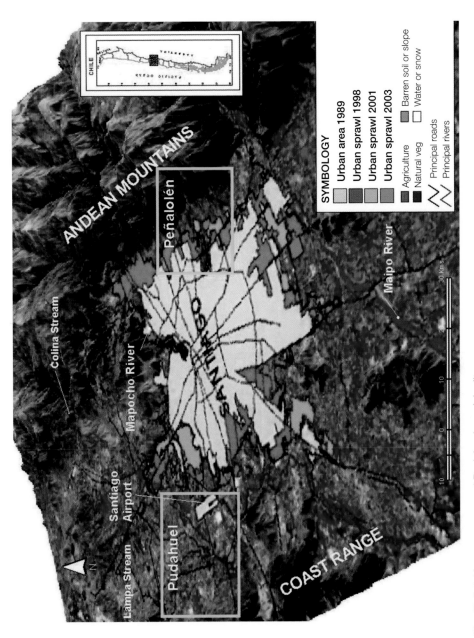

Figure 9.1 The Peñalolén commune and Pudahuel municipality

Image supplied by author

Figure 9.7   Peñalolén medium-class condominium

Photo by M. Molina and C. Moscoso

Figure 9.8   Peñalolén social basic dwelling

Photo by M. Molina and C. Moscoso

Figure 9.9   Macul stream in the last flood, 27 August 2005

Figure 9.10   Nido de Aguila stream in the last flood, 27 August 2005

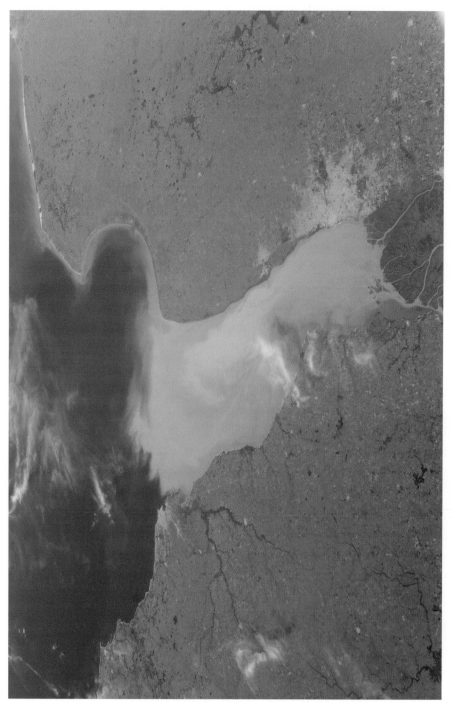

Figure 10.1 Satellite image of the Buenos Aires area

Source: NASA Johnson Space Center – Earth Sciences and Image Analysis (NASA-JSC-ES&IA) (2003)

Figure 11.8 Soil erosion in the freguesia of Ceira, February 2006

Photo by J. Duarte

Figure 11.9 Destruction of terraces due to fire and the abandonment of traditional agricultural methods in the freguesia of Ceira, February 2006

Photo by J. Duarte

Figure 11.10 Destruction of terrace retaining walls due to fire and the abandonment of traditional agricultural methods in the freguesia of Ceira, February 2006

Figure 11.11   Debris flow in the freguesia of Ceira, February 2006

Figure 12.1 Students measure the cut trees

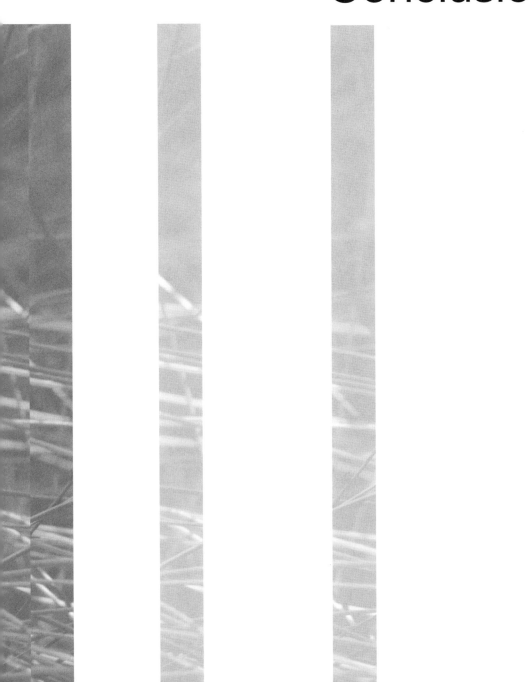

Part C: Conclusion

13 | Our capacity for sustainable nation-building: Do we flourish or flounder?

Margaret Robertson

We face a critical period in our global history. The experts on climate change accept this as fact. Likewise, many nations accept their responsibility to make the changes necessary, such as enacting legislation, to halt inappropriate behaviours by industry, commercial enterprises and domestic households. However, reality is that accepting the significance of the rhetoric is one thing. A commitment to legislative responses and transformed behaviours is another. The real test comes in the quality of the legislation and the support mechanism that allow it to be enacted within a short time frame.

Nation-building strategies are facing this test around the globe. As many observers note, the effectiveness of the Kyoto Protocol as the definitive benchmark for nations has had variable success. Why? Much of this examination needs to take place in the context of 'who holds the power'. In an exploration of the meaning of power in the global age, Ulrich Beck (2005) suggests the need to delve into new ways of thinking. Reflecting on Bourdieu's distinction between economic, social and economic capital, he sees limitations in the underlying assumptions. European philosophers of postmodernist thinking, while paving the way for rethinking issues in the latter half of the twentieth century, have lost ground as we move from western perspectives to postcolonial visions and beyond. Global influence is shifting east. In Beck's view, we therefore need to introduce the 'meta-power game between state, capital and social movements' (Beck 2005, p. 240). That is 'legitimatory capital' or that which 'enables advocatory networks to ... establish new concerns and categories in opposition to much more powerful organisations and governments' (pp. 240–1).

Interestingly, at the time of writing, this process is influencing the 2007 federal election outcome in Australia. Following a year of record levels of low rainfall and high temperatures, the scientific warning signs of global change are impacting on voters through enforced water management policies at local and state levels of administration. The legitimacy and profile of climate change scientists and global spokespersons, such as high-profile former US Vice-President Al Gore, are running at unprecedented high levels. The personal 'pain' is permitting voters to capture the panoptic first-hand.

According to Beck, fostering change for sustainable futures requires '*internationalization* of the nation-state' (Beck 2005, p. 261, original emphasis). He cites examples of the effectiveness of non-government organisations in forcing parochial-thinking governments to look towards the goal of 'establishing a world order based on civil society' (p. 288). This notion of global 'cosmopolitanism', as Beck describes it, would eliminate boundaries and barriers between people. It would be an ideal world, perhaps, but sadly more like an imagined fantasy, since it seems difficult to imagine the human psyche acquiescing to such sophisticated negotiations. The destiny of the world's nations seems to be the irrepressible hegemonic pushes for power by political factions—unless some extraordinary natural disaster precipitates a new readiness to adopt 'cosmopolitan' levels of maturity in negotiations.

'Seeing differently'—the power of education

In her book *Structure, agency and the internal conversation*, sociologist Margaret Archer (2003) provides a simplified means for individual students and their teachers to grasp the complexities of interpersonal relationships. If we, as educators, are to prepare young scientists growing up in this complex world of frictional and conflicting interests, then we need to pass on some guiding principles. Archer refers to the need to acknowledge and allow for the 'private' life of the individual. Constraints on personal aspirations can lead to disharmony and unwanted reactions. However, legitimising introspection and our need to internalise ideas within the framework of our existing knowledge is a way to develop the conversation towards positive outcomes. She describes this subjectivity as recognition of '*a personal internal property*, with a *first-person subjective ontology*' (Archer 2003, p. 15, original emphasis). Structures may give some element of objectivity in the context of people and place, but the agency—and who acts as the agent—place the action firmly in the realm of the subjective. In summary, Archer argues that 'how people reflexively deliberate upon what to do in the light of their personal concerns has to form a part of mediatory account' (p. 15). Understanding this process of introspection and its context of personal seems to be fundamental to the process of learning to respond in collective and constructive ways.

Our contexts are never simple, as the case studies presented in this book demonstrate. Each author discusses complex social and cultural interactions with nature. Layers of imagined worlds are likely to be juxtaposed in any specific location. It is helpful, then, to refer back to the place where the conversation that led to this book and the various projects originated. The place is Siberia and the following text is from Colin Thubron's book *In Siberia*.

The city of Khabarovsk trickles along three ridges where the Amur and Ussuri rivers unite and the border with China veers south. It was founded in 1858 by the belligerent governor Muravieb-Amursky during his push to the east, and was named after his distant predecessor, the merchant-adventurer Khaborov, who had savaged his way down the Amur two centuries before ... Among the Ladas and Zhingulis the streets are scattered with Toyotas and Nissans: Japan is only three hundred miles away. I idled here with a sense of being in Europe ... I took a bus to the chief market. It was swarming with Chinese vendors. They had arrived overland or by river-boat from Harbin ... A shadow land of pirated logos and yearned-for cities pervaded their wares: New York, Paris, Milan, Fake Reebok and Adidas shamelessly abounded ... This alien presence is nothing new. By the end of the nineteenth century every town in south-east Siberia had a burgeoning Chinese quarter. (Thubron 2000, p. 244)

The coming together of east and west in this far north-eastern part of Russia seems symbolic of the melting pot of ideas, and sociocultural and economic perspectives, that exists in any location we may choose. Siberia has been populated by forced migrations, courageous individuals able to withstand the extreme weather conditions, and by waves of travellers through the centuries. Ecotourism has opened the borders to this seemingly end-of-the-line part of the world. But as Thubron points out, the legacy of Soviet rule remains strong—travellers must register at their hotels, and relinquish their passports for inspection.

Agency and structure are clear. However, as our travelling group of scientists was able to experience first-hand during the summer of 2002,[1] young people in schools and university are looking outwards for answers and seeking resources to develop their skills and knowledge. We met students who were acutely conscious of their remoteness from mainstream education facilities, but who had clear visions and were able to articulate the shared concerns for the environment and how best to manage sustainable practices within their context. Educational support is slim but perhaps this book is one way of bringing the much-needed learning tools to the next generation to further develop their skills and knowledge.

Moving forward

In the concluding chapter of her book *Sustainable landscapes and lifeways*, Anne Buttimer (2001) suggests that the way forward for progress in sustainable development is to include those voices which are less often heard. In her analyses of European contexts, she comments: 'Involvement of voices from lived experience can provide a vital complement to observations based on archival and landscape records' (Buttimer 2001, p. 384). She further observes that 'Analyses of landscape morphology or aesthetics yield a relatively opaque description of sustainable development' (p. 384), meaning that local landscape practices derive from many historical influences and evolve over time, and landscape morphology is only one part of the picture. Understanding local layers of thinking is the vital step for national and global agencies who aim to bring about transformative behaviours. The goal to achieve sustainable practices should remain intact. The process for achieving that goal will need to be adapted to suit local people, their lifestyles and their perceived place in the future of their lands.

Our book exemplifies this principle. There were to be nine completed case studies, but Morris Chauke in South Africa faced considerable local delays—not because of an urgent need to solve the local water problem, and not because of the will to bring this about. His lack of resources has been his biggest enemy. The central issue is linked to capacity, as well and financial and time constraints. A major concern for the authors is that their projects will provide support for the poorest communities of the world and encouragement for local students. As a group we share this responsibility and concern for our colleague. The planning for the South African local projects is complete (see Appendix), and it may still move forward. The Celiographers are committed to support their colleague and provide assistance where possible.

So, the road ahead may not be easy. Obstacles are inevitable and may prove too large to circumvent. Nevertheless, shared conversations are themselves an important starting point in the quest for greater knowledge and skills to assist with local custodianship of the planet—regardless of local contexts and circumstances.

In the case studies presented here, we highlight the inevitable environmental tensions that occur with urbanisation and concentrations of people into very large cities. Water scarcity and the management of limited existing resources have been highlighted as a major issue by WU Shaohong of Beijing (China) and Shyam Asolekar of Mumbai (India). In Argentina's Buenos Aires, Gabriel Bautista widens the urban problem to the humanitarian issues including access to food, water and shelter. Charlchai Tanavud's analysis of disaster management in the aftermath of the 2004 South-East Asian tsunami shows how anticipation of events by local people can help with preparations; this is also true in the management of forest fires, as represented in Manuela Ferreira's case study of Portugal. Álvaro Sánchez-Crispín's case study of the Tres Palos lagoon area of Mexico and Niko Beruchashvili's guiding case study of teaching for sustainable development demonstrate practical ways of taking the next step in educating students about sustainable development. This is also the case in Chile, where Hugo Romero and Alexis Vásquez consider how to cope with urban sprawl and teach for improved understandings.

Together, these case studies offer hope for students and their teachers. The Celiographers hope that their efforts will inspire many more local projects that can add to this new 'cosmopolitanism' and reinforce the value of international collaboration. Such local projects provide material evidence of how the application of leading-edge thinking on global issues is having an impact at local levels. Being part of this project and any that may follow is a privilege and inspiration for all who may be trying to imagine a better world and an environmentally safe planet.

Bibliography

Archer, M 2003, *Structure, agency and the internal conversation*, Cambridge University Press, Cambridge.

Beck, U 2005, *Power in the global age*, Polity Press, Cambridge.

Buttimer, A (ed.) 2001, *Sustainable landscapes and lifeways*, Cork University Press, Cork.

Thubron, C 2001, *In Siberia*, Penguin Books, London.

Notes

1   In 2002, led by Professor Nikita Glazovsky, the International Geographical Union held a regional meeting in the city of Barnaul and surrounds. Here the proposal for developing the case studies included in this book took place.

Appendix

South Africa projects—Morris Chauke

	Habitat/ Geographical setting	Threats/ Processes/ Weaknesses	Nature of intervention/ Opportunities	Stakeholders engaged in action	Pedagogic tools	Research topic/ Location
South Africa—1	Urban and rural settlements	• Water shortage • Poor water management	• Preserving water resources	• Learners • General public • Local government • Traditional leaders	• Learner-group activities • Interviews • Questionnaires	Development of basic water management strategies *South Africa: Limpopo: Makhado rural and urban areas*
South Africa—2	Rural community	• Land degradation • Water pollution • Invasion of alien plants	• Preserving ground-water resources • Protection of biomes	• General public • Local government department • Traditional leaders and doctors • Educators and learners	• Field visits • Assignments • Discussions • Lectures • Interviews	Negative impact of human activities on the sustain-ability of local biodiversity areas *South Africa: Vhembe District*
South Africa—3	Rural community	• Decline of local communal food production • Poverty and malnutrition	• Finding alternative means of food production. • Investigating human links with climate change	• The general public • University students • School learners • Local authorities • Industrialists	• Lectures • Assignments • Group discussion and tasks	The effect of climatic change in sustainable rural communal farming *South Africa: Soutpansberg West communities*

Subject index

Abler, Ron   vii, xi, 142
Acapulco   126, 131–42
acting locally, thinking globally   6
action learning   8, 101, 116
active citizenship   205
Activity Theory   xiv, 33–4
aerial photographs   154, 216–17
aesthetics   24, 43, 129, 212, 229
agency   26–8, 33, 34, 225, 228–9
   human   10
agriculture   32, 53, 128
   Asian region   64, 83, 97
   Latin America   135
   Portugal   197, 200–2
analytical skills   13
aquaculture   66, 67, 72, 153
artwork and posters   75–6, 94, 181–3, 185,
   187
Asociación Latinoamericana de Integración
   (ALADI)   27
Association for South East Asian Nations
   (ASEAN)   27
Association of American Geographers   vi
Atacama Desert   152
Australian Academy of Science   xi

belief/s   10–11, 12, 23, 25, 34
Bellezza, Giuliano   viii, xi
biodiversity   4, 31, 42, 49, 50, 57, 169, 176
bioresources   55
British Council for Geographic Education   vii
Brundtland Commission   39, 43, 127
Buenos Aires   149, 179–90
Buttimer, Anne   v–viii, xi, 229

capacity-building   v, 3–4, 9, 28, 34, 227–31
   disaster management   63–76
cartographical analysis   16, 154
Ceira   192, 200–7
Celiographers, The   viii, 7, 9–11, 230
children   179, 181, 183–7, 204
citizenship   47, 187, 204, 205
clean water   23, 29, 96
climate change   31–2, 36, 54, 55, 227
Club of Rome   38
collaboration   v, vii, 5, 25, 205–6, 230
Communities of Practice   xiii, xiv, 6, 206
community
   management   179–90
   participation   4, 32, 113, 117, 128, 140,
      142, 168, 180, 189
   partnerships   8
   planning   42–3, 89, 207
Complexity Theory   34
Council of Europe   44
critical thinking   11, 204

deforestation   50, 55, 66, 148, 162, 222
Delhi   105, 109–11, 113, 116
democracy   17, 23, 40
desertification   49–50, 55, 64, 202
deserts   209
dialogue   vii, xiii, 5, 25, 30, 171, 189
disasters   51, 52, 67, 68, 70–2, 74
   disaster management   63–76
dwelling   30, 31

earthquakes   52, 65, 68, 70
Earth Summit   39

eco-feminism   30
ecological equality   57
ecology   vii, 57, 59, 102–4, 176, 221–2
economic growth   11, 58, 138, 152, 213–14
economic power   22
economies, old   25, 59
ecosystem(s)   32, 50, 57, 163, 169
    aquatic   53, 71, 101, 102–4, 137
    coastal   32, 67, 175
    natural   67, 104, 149, 169, 182
    urban   103–4, 146, 182
education   160, 168, 199
Education for Sustainable Development
        (ESD)   15, 40, 41–2, 204, 207
El Niño   4, 51, 168
emissions   51, 53, 149
environmental ethics   45, 47, 56–7, 58
environmental management   xiii, 3, 24, 31,
        151–2, 163–8, 170
erosion   50, 63–4, 66, 68, 69, 102, 104, 148,
        151, 168, 172–97, 201–2
Europe   25, 44–7, 191–223

fieldwork   43, 154, 175–6, 180, 181, 182,
        183, 187–8, 206, 214, 216
fire   191–207
fishing   56, 135, 136, 138–141, 153
floods/flooding   52, 65, 67, 70–2, 155, 162,
        169, 172, 180–1, 184, 187–90
fluvial-lagoon   126–42
Food and Agriculture Organization
        (FAO)   51
forest fires   191–207
forestry   4, 153, 218, 219–221
forests   46, 49, 51, 65, 67, 72, 137, 155–6,
        166, 191–207, 209, 212, 214, 215, 218,
        219
    fuel mass   198
    management   112, 199, 212, 220–1
Foundation for Environmental Education
        (FEE)   46, 47

Gandhi, Mahatma   15, 22–3, 34
Geographical Information Systems (GIS)   71,
        73, 75–6, 115, 153
geopolitical differences   xiii
Georgia   209–23
geospatial knowledge   11

Glazovsky, Nikita (dec.)   vi–vii, viii
globalisation   vi, 17, 27, 148, 152, 184,
        204–5
global warming   22, 23, 32, 183
governance   10, 27, 33, 34, 150–1
governmental, inter-   26, 27
green belts   85–6, 92–3, 103, 156, 161–2,
        166, 168, 184
green cover   93, 102
greenhouse effect   50, 51–2, 53, 54, 191
green politics   26
groundwater   79, 81–4, 105, 106, 110–11,
        112–13, 116, 118, 150

hazards, natural   76, 146, 150, 168, 172,
        176
health, public   17, 40, 101, 149, 156, 160,
        166, 168, 175
Heidegger, Martin   30–1
herbicides   53–4, 55
Home of Geography   vi, vii, 6
horticulture   32
hotels   93–5, 116
human agency   10
hydrological cycle   162, 188, 189

Indian Ocean   71
Indigenous Peoples of Africa Co-ordinating
        Committee (IPACC)   27–8
Industrial Revolution   49, 51
intercultural learning   5, 11
Intergovernmental Panel on Climate Change
        (IPCC)   32
International Council for Science (ICSU)
        v, vii
International Geographical Union (IGU)
        v–viii, 29
    Commission on Geographical
            Education   vi, vii
irrigation   85–6, 93, 100, 106–7, 155–6, 162,
        172

knowledge, oral   76
Kyoto Protocol   227

lagoon   126–42
land degradation   50, 55
La Niña   51, 172

Leadership for Environment and Development
     (LEAD)   vi, 4
leisure, productive   212
local area networks   vii, 4, 7–8
local communities   3, 23, 76, 154, 169–70,
     187, 189

management   32, 39, 86, 103, 146
     municipal   154–68, 170
Mandela, Nelson   22–3, 29, 34
Mediterranean vegetation   150, 152, 173,
     192
memories   188
Mexico City   131
middle class   26, 150, 160, 185, 203
Millennium Declaration   40
mining   55, 128, 135, 153
Moken   76
Montevideo   149
Mumbai   100–21

national parks   104, 132, 210–22
nation-building   227
natural disasters   52, 65, 67–8, 70, 74, 75, 76
     see also natural hazards
natural hazards   66, 69, 71–3, 76, 149–51,
     162, 168, 172
nature, and society   49, 126, 127, 184
New Delhi   110

Organisation for Economic Co-operation and
     Development (OECD)   27
overpopulation   56, 149
ozone   51–2
     ozone depletion   51–2, 53

pesticides   53–4, 55, 141
place, sense of   5, 129, 183, 190
place-space connections   11
policy, economic   57–8, 59, 128, 161
pollution   38, 100, 128, 146, 150–1, 175
     air   53, 55, 149, 174
     coastal   126–42
     marine   53, 55
     soil   53–4, 55
     water   53, 55, 103, 110, 126–42, 148, 149
population migration   108, 117, 129, 146,
     147, 197, 209

postcolonialism   30, 227
poverty   33, 40, 147, 152
     rural   66, 117
     urban   156–161, 165
Prince of Songkhla University   74–5
problem-solving   13, 43, 171, 204, 205–6
programmed knowledge   8
project management   6
public health   17, 40, 101, 149, 156, 160,
     166, 168, 175

rain   187–90
     harvesting   100–21
     recycling   92–3, 96–7
recycling   26, 58
reforestation   191, 197
remote sensing   76, 115
resilience   76
Rio — Agenda 21   39–40
Rockefeller Foundation, The — Leadership
     for Environment and Development
     (LEAD)   vi
Rome   v, vi, 6–8, 11, 20, 179, 210
round-table discussion   210, 215, 221
rubber   65–7, 70
run-off   80, 93, 102, 105, 109, 117–19, 149,
     162, 202
Russian Academy of Science   xi

Sahara Desert   55
sanitation   11, 158, 166
Santiago   146–76
satellite imagery   65, 71, 115
science students   4, 8–9
sewage   29, 53, 135, 136, 140
     treated   102, 110
sewerage   141, 158
skill building   115–6
skin cancer   52
snow   51, 117, 150
social capital   10
social integration   17, 151, 162
social justice and equity   15, 17, 112, 204
social responsibility   26
socioeconomic environment   156, 158
soil degradation   63
soil erosion   50, 54, 64, 68, 102, 148, 168,
     172, 176, 201

South Africa   23, 29, 40, 230, 232
Soviet Union   212, 229
space   11, 17, 24, 25, 26, 29, 34, 153
spatial orientation   24
storytelling   183, 185
    oral histories   188
student training   7–8, 205, 215–223
subjectivity   228
sustainable lifestyles   15, 23
systems approach   24, 33–4, 185, 190

Tbilisi   44, 209–223
    national park   211–13
The World Conservation Union (IUCN)   50
tides   52, 104
tourism   67, 72, 126–42, 197, 210, 211, 214
    ecotourism   128, 130, 138, 229
    geotourism   129
training   7–8, 44, 75, 199, 215–23
transportation   180–3
tsunamis   65, 68, 70–2, 74–6, 230
typhoons   52, 65, 67

underground water   93, 105
United Nations Conference on Environment
    and Development   39
United Nations Decade of Education for
    Sustainable Development   15, 41–2,
    204, 207
United Nations Educational, Scientific and
    Cultural Organisation (UNESCO)   19,
    28, 44

United Nations Environment Programme
    (UNEP)   44, 50
urbanisation   105, 146–7, 148, 151, 156–7,
    162, 173, 197, 230
    urban sprawl   146–76, 179, 181, 184, 187
urban planning   161
urban-rural interface   180

Vygotsky, Lev   33

waste   53–4, 67, 104, 134–5, 136–7, 141,
    156–7, 166, 167
    domestic   156–7, 166
wastewater   87, 96, 104, 135, 141
water
    harvesting   100–21
    management   79–98, 100–21, 227
    recycling   86, 92, 93
    scarcity   55, 87, 98, 101–2, 108
    see also clean water, groundwater, lagoon,
        rain, run-off, underground water,
        wastewater
water consumption   79, 82–7, 91–4, 97, 98
water pollution   53, 55, 103, 110, 126–42,
    148, 149
water rights   153
water saving   79–98, 100–21
watersheds   65–7, 102, 148, 155, 169–70,
    172, 179, 188
wetlands   103–3
World Commission on Environment and
    Development   39, 49, 127

Author index

Acosta, R   138, *143*

Aravena, D   155, *176*

Arbhabhirama, A, Phantumvanit, D, Elkington, J & Ingkasuwan, P   63, 65, 77

Arceivala & Asolekar   *121*

Archer, M   228, *231*

Arnold, C & Gibbons, J   162, *177*

Arredondo, J, Ponce, J, Luna, C, Coronel, C & Palacios, C   135, 137, *143*

Arredondo, J, Quiñónez, R, Luna, C, Coronel, C & Palacios, C   137, *143*

Asolekar, S   *121*

Asolekar, S & Gopichandran, R   *121*

Axford, B   *35*

Banderas, A   *143*

Barnes, K, Morgan, J & Roberge, M   148, 162, *177*

Beck, U   227–8, *231*

Beruchashvili, N   *223*

Beruchashvili, N, Chauke, M & Sánchez-Crispin, A   *223*

Beruchashvili, N & Zhuchkova, V   *223*

Brown, L & Kane, H   *59*

Buell, L   24, *35*

Butler, R   128, *143*

Buttimer, A   32, *35*, 229, *231*

Cabral, S   40, *47*

Carrega, P   198, *207*

Catarino, V   197–9, *207*

Caviares, A   175, *177*

Corvalán, P, Kovacic, I & Muñoz, O   162, *177*

DeGraff, JV   65, *77*

Delaney, D   26, *35*

Dewey, J   29, *35*

Dyball, R, Beavis, S & Kaufman, S   34, *35*

Eidse, F & Sichel, N   31, *35*

Engeström, Y   33, *35*

Espinoza, G, Valenzuela, F, Jure, J, Toledo, F, Praus, S & Pisan, P   163, 165, *177*

Fan, B   *59*

Ferreira, M   43, *47*, 204, *208*

Flavin, C & Lenssen, N   *59*

Freire, P   30, *35*

García, E   *143*

Gupta, G   *122*

Hall, CM & Lew, AA   *143*

Han, G & Wang, L   *98*

Harvey, D   30, 31, *35*

Hay, P   30, *35*

Honey, M   128, *143*

Houtsonen, L   127, *143*

Huang, W & Niu, Y   *59*

Hui, S, Xie S & Zhang, S   *98*

Keen, M, Brown, VA & Dyball, R   31, *35*

Kemmis, D   187, *190*

Lanza, A   126, *144*

LeFebvre, H   29, *36*

Li, R   *60*

Li, X & Zhang, W   *98*

Liu, Y, & Li, X   50, 51, 56, 60
Liu, Y & Zhou, H   60
Lourenço, L   208
Lu, Y   60

Ma, Y   51, 52, 60
Macnaghten, P & Urry, J   24, 26, 36
Massey, D   34, 36
Meadows, DH, Meadows, DI & Randers,
    J   39, 47
Meadows, DH, Meadows, DI, Randers, J &
    Beherens III, W   38, 47
Miranda, B, Alexandre, F & Ferreira M   144,
    204, 208
Munasinghe, M   19, 20

Neville, B   30, 36
Nisbett, RE   25, 36

Pile, S & Thrift, N   10, 20
Pita, L, Cruz, M, Ribeiro, L, Palheiro, P &
    Viegas, D   198, 208
Pittock, AB   32, 36
Postel, S   60
Propin, E & Casado, J   134, 144
Pugh, C   33, 36
Purvis, M & Grainger, A   20, 144

Qian, Y   99

Rajabu, K   124
Rau, JL   70, 77
Rebelo, F   192, 208
Rego, C   197, 208
Robertson, M, Webb, I & Fluck, A   8, 20
Romero, H   150, 172, 177
Romero, H, Ihl, M, Rivera, A, Zalazar, P &
    Azocar, P   149, 177
Romero, H & Vásquez, A   149, 161, 162,
    177–8
Romero, H, Vásquez, A & Ordenes, F   149,
    177
Roseland, M   10, 20
Rougurie, G & Beroutchachvili, N   223

Sánchez-Crispín, A & Propin, E   134, 144
Schücking, H & Anderson, P   32, 36
Shiva, V   30, 36
Slater, F   43–4, 47
Stokes, E, Edge, A & West, A   45, 48

Tanavud, C   70, 76, 77
Tanavud, C, Bennui, A & Sansena, T   77
Tanavud, C, Yongchalermchai, C & Bennui,
    A   65–6, 72, 77
Tanavud, C, Yongchalermchai, C, Bennui, A
    & Densrisereekul, O   65, 67, 70, 77
Tanavud, C, Yongchalermchai, C, Bennui, A
    & Navanugraha, C   70, 73, 77
Tanavud, C, Yongchalermchai, C &
    Densrisereekul, O   66–7, 70–1, 77–8
Tanavud, C, Yongchalermchai, C,
    Navanugraha, C & Bennui, A   68–9, 78
Thubron, C   228–9, 231
Trudgill, S   29, 36
Tuan, YF   24, 36, 190

Usher, AD   66, 78

Vargas, S, López, E & Romero, R   144

Wan, Y & Jin, D   99
Wang, R   99
Wang, T   60
Wheeler, SM & Beatley, T   39, 48
Wilbanks, TJ   16, 20

Yan, C   99
Yang, Q, Zheng, D & Wu, S   60
Yokoyama-Kano, A   137, 144
Young, L & Hamshire, J   127, 144

Zandbergen, P, Schreier, H, Brown, S, Hall, K
    & Bestbier, R   155, 178
Zhang, X & Zhang, Y   60
Zheng, G & Lu, J   80–1, 99
Zhu, Z & Liu, S   60
Zhu, Z & Shao, Z   82, 99
Zuo, D   60